Richard Lacey is Professor of Clinical Microbiology at Leeds University. He was educated at Jesus College, Cambridge, taking a first degree in medicine, before turning to clinical microbiology with a Ph.D. from the faculty of medicine at the University of Bristol.

He acts as a consultant to the World Health Organization and was advisor to the EEC 'Flair' project on Food Safety Research Grants in 1990. He was also a member of the Veterinary Products Committee in the Ministry of Agriculture, Fisheries and Food from 1986 to 1989.

He has received numerous awards, including the Evian Health Prize for Medicine, the Caroline Walker Award for Science, and an award from the Campaign for Freedom of Information.

He has given over 150 television interviews and 400 radio interviews world-wide, and was a columnist for the *Yorkshire Evening Post* for three years. He was scientific advisor to the highly acclaimed BBC food thriller, *Natural Lies*, in 1992.

You are what you eat – or are you?

What is in food? Where does it come from?

Richard Lacey takes the reader on a culinary exploration into the world of food. Blending science and humour, he stimulates us to question the future and to think about the nature of what we eat and where it comes from. Richard Lacey is on the side of the consumer, you and me, as he reveals the sinister side of food production and the dangers lurking in the kitchen.

The reader is served up with a feast of practical tips on the handling of food.

But food is FUN too! Our taste buds work overtime as we are shown how to enjoy food that is delicious, healthy and safe.

The overall message is: enjoy your food but be aware of the dangers and take care. As you read you will laugh, wince and learn about FOOD.

Also of interest in popular science

Remarkable Discoveries!
FRANK ASHALL
An inventor in the Garden of Eden
ERIC LAITHWAITE
The Outer Reaches of Life
JOHN POSTGATE
Prometheus Bound
JOHN ZIMAN

RICHARD W. LACEY

Hard to Swallow

A brief history of food

CAMBRIDGE
UNIVERSITY PRESS

CAMBRIDGE UNIVERSITY PRESS
Cambridge, New York, Melbourne, Madrid, Cape Town, Singapore, São Paulo

Cambridge University Press
The Edinburgh Building, Cambridge CB2 8RU, UK

Published in the United States of America by Cambridge University Press, New York

www.cambridge.org
Information on this title: www.cambridge.org/9780521440011

First published 1994
Reprinted 1994
This digitally printed version 2008

A catalogue record for this publication is available from the British Library

Library of Congress Cataloguing in Publication data

Lacey, Richard (Richard Westgarth)
Hard to swallow : a brief history of food / Richard W. Lacey.
p. cm.
Includes index.
ISBN 0 521 44001 7 hardback
1. Food. 2. Food industry and trade. 3. Food contamination.
4. Nutrition. I. Title.
TX353.L23 1994
641.3—dc20 93–20808 CIP

ISBN 978-0-521-44001-1 hardback
ISBN 978-0-521-06494-1 paperback

To my Mother and Father

Contents

Preface

At this stage, I will resist the temptation of presenting a potted précis of each chapter; nor will I summarise all the main conclusions. But I will explain what I have tried to do – that is, provide the essential facts on food; and the central purpose of each chapter is to tell you about different aspects of what has come to be called the food chain. This is an appropriate phrase, as it suggests the possibility of a few weak links in the provision of our food.

But we will not dwell on the problems, and I hope that the section dealing with the pleasure of eating will counter the horror stories in Chapter 9.

I wrote this book because I wanted to, and I hope you do not judge it to be just another book promoting particular types of food. The environmental theme is dominant. We are simply wasting the world's resources unnecessarily on the way we produce food at present.

Thinking of a plan to put matters right produced a feeling nothing short of despair in me. If you want to stay cheerful, stop at Chapter 17, and enjoy some of my mother's favourite recipes.

Some readers might find certain sections rather flippant for a work that includes some serious scientific and medical facts. But surely there is something ridiculous about much of what – and how – we eat and drink? Hopefully, you will easily identify the light-hearted asides, whose function is intended to provoke thought more than anything.

Another purpose of this book is that it will stimulate you to question in the future what exactly you are eating – what is in it, where does it come from, what advertising slogans really mean and, if it is meat, how the animal was reared and slaughtered.

It is hoped that this book will appeal to some vegetarians, or to those people who have not made the decision to go all the way, because of the proposal that many of our environmental problems have been caused by our habit of eating too much meat.

This is necessarily written from my UK experience but hopefully most of the substance is also for consumption in North America; I suspect that North American readers will be appalled at the UK's food poisoning record. However, food from both sides of the Atlantic does show many features in common, and it may be rather unwise to generalise. At the time of writing, the $64,000 questions are whether Mad Cow Disease (BSE) will disappear from the UK herds as if by magic, and whether it has already infected people. But don't let this put you off North American beef completely. However, the potential knock-on effects from this disease for world farming are very great and, however unpalatable, they must be faced with honesty.

The chapter on the ideal diet is intended to be reassuring. If you are old enough to read this, some damage to your arteries will probably already have occurred! But you can help the next generation. It is definitely not my intention to fan the flames of cholesterol-phobia now sweeping the world: quite the reverse. I have tried to present the facts relevant to cholesterol. They are sometimes contradictory and confusing. In preparing readers for these, the following summarises the position as I see it. In most people, the amount of cholesterol in the blood is not due simply to the eating of cholesterol itself, but due to diet or lifestyle. Saturated animal fats and hydrogenated oils *do* make your blood cholesterol rise. The purpose of the well-balanced diet is to stop this happening, and is positively encouraged. But a raised blood cholesterol is but one of the many consequences of certain lifestyles, and it has not been proven that attempts to lower it by, say, drugs in middle age, does much if any good for your long-term health. Claims that processed food might be beneficial because it is low in cholesterol can be completely misleading if such food is high in saturated fats or hydrogenated oils.

For most people, the message is simple. Ignore the cholesterol propaganda, don't have your blood tested for cholesterol and eat an enjoyable and varied diet as suggested here. The real difficulty is to get the amount of food you need right.

The text is hopefully organised in a logical way, beginning at the beginning with crops, going on to wild meat and then intensively farmed meat and salmon (you may be disturbed about this), then to the various ways we endeavour to ruin certain types of whole food by processing.

Then we touch on disasters before getting onto the serious business of eating food, and enjoying it, finally coming to the depressing end. Overall, I have tried to be fair to the food industry. Some products are excellent, others are not.

Some commentators may question the authority I have to write about a subject so important and all embracing to our society. Critics will point out that I am not a 'Food Microbiologist'. True. I am proud of this. I don't work for the food industry. But I think I have one or two tentative qualifications, such as a career in medicine with special interests in chemistry, cholesterol, children, microbiology and infective problems of food. In addition, I have been a lifelong ecologist.

I would like to thank those people or organisations mentioned in the text for permission to be so quoted and, in particular, Chris Hardisty, for her recipe for tofu, and my mother (a sprightly lady whose writing is infinitely better than her doctor son's) for the recipes in Chapter 17. Incidentally, I was brought up on these recipes, interjected between the most dreadful food at boarding school. No wonder I never wanted to return from vacations!

I would like to thank Mr. Andrew West for the drawings, which may well convey a more cogent message than the rambling text.

I should also like to thank my wife Fionna for grammatical and culinary help, and my secretary, Mrs Hilary Mobbs, who must be empowered with astrological insight to understand my writing!

PART I
Farming

1
Plant crops

The diet of our ancestors

When did farming begin? Or, if you like, when did *we* begin?
Compared to the time taken for us to evolve from lower primates,
the answer for farming is very recently indeed. So it is not surprising that in just the last few years we have realised that some
serious mistakes have been made in the way we produce our
food. But we still seem reluctant to test new farming methods
adequately before rushing into massive commercial exploitation
of them.

Our earliest ancestor is now thought to be a fairly ordinary and
smallish type of monkey living in the trees of a luxuriant rain
forest. This was 30 million years ago, and the territory is now
barren desert in North Africa. We know from the teeth of this
animal that it fed mainly on vegetation, with dagger-like canines
for breaking the tough casings of fruit.

We can now assume that early primates gathered their food
which was available all the year round. Each animal would concentrate on the need to feed itself, apart from the mother nourishing
her young. There would be no need to store food and there would
be little sociable feeding. No dinner parties!

By around 20 million years ago our ancestors were venturing
outside forests, but were still predominantly vegetarian.

Thirteen million years ago one such ancestor, called *Dryo-
pithecus*, had migrated to the European forests and evolved into
gorilla-like animals, so by 10 million years ago the teeth were
getting stronger and larger and this enabled tougher food to be

eaten. This meant that some of the diet still essentially of vegetable origin could be stored in a dried form, so that these primates could survive in drier territory, not needing the constant
availability of edible parts of trees.

Exactly when our ancestors began to eat meat is not known, but
the Southern Ape of 4 million years ago was still predominantly
vegetarian. However, *Homo habilis* 2 million years ago included
meat in its diet. This would be obtained largely from dead mammals or birds.

It was not until just around 250,000–500,000 years ago that
Homo erectus is thought to have purposefully killed and cooked
animals such as deer and rhinoceros. Cooking might suggest
some social organisation, but there would then have been no
attempt to cultivate plants or rear animals specifically for food.

Up to about 4000 BC our ancestors, now sufficiently intelligent
to be described as *Homo sapiens*, continued to gather edible vegetation and fruits and hunt animals. But since that date 'we' began
to grow crops and to control animals and birds for food. Most
evidence for this exists in the Middle and Far East. We were
soon to learn to improve the desirability and yield of plant crops,
and also 'improve' animals by selection and hybridisation (this
means cross-fertilising two similar species to produce a range of
resulting hybrids, the best of which are then grown on).

We were also to learn to exploit microorganisms – such as
bacteria and yeasts – although for several thousand years we
would not understand exactly why the processes involving them
were effective. Brewing and the making of cheese are part of our
ancient history, as is the use of early food preservatives – drying
and the use of salt and sugar.

But improvements in crops and food animals were slow, and
the technological changes in our food production over the last
70–80 years have been much more dramatic than in the previous
30 million. For example, the practice of artificial insemination
has revolutionised animal husbandry, both in breeding and in
rearing. A bull, boar or a ram are rare sights today! Genetic
engineering, and the use of artificial fertilisers and chemicals
against pests, have enabled the world to produce dramatically
more food than it used to from the same space. Indeed the world

production of food increased at an annual rate of around 3–5% during the 1980s.

The world population has also increased, and we are now acutely aware that despite increases in food production, in many countries starvation persists and we increasingly recognise diseases associated with eating either too much, or the wrong type of food.

The effect of this dramatic increase in 'efficiency' in agriculture on the world's environment is already producing tragic and potentially long-term devastating consequences.

Crops today – grassland

Apart from new buildings and factories, most of our environment seems to present to its human inhabitants a sense of permanence or timelessness. But the fields, hedges and lines of trees in the country are nearly all constructed by us. The natural vegetation of most of North America and Europe was forest, mainly conifers in the north and broadleaved trees in the south. Moreover, a great deal of the deforestation has occurred in the last few centuries and soil fertility already has to be maintained by artificial chemicals.

Most of the grassland in the world is therefore man-made, and the types of grasses that now provide the grazing for cattle and sheep would originally have had to struggle to survive within the hearts of the forests, but would have flourished at their edges – as the forests met rivers, cliffs or clearings.

There is still some natural grassland in the world such as the exposed inclement southern part of South America and Eurasia, and also high up in the mountains, where trees have given up the unequal struggle against the cold.

Grasses are a crop in themselves and keen gardeners are aware of the different types of lawn seed. Farmers are too, and use grasses mixed with for example clover, to increase the nitrogen content of the soil.

Where tropical forests have been felled to make way for grazing fields, the natural grass is different from that of most European and some North American species, tending to be coarser. Fine perennial temperate grass in Brazil is referred to as English grass!

Grassland is therefore a specialised crop in its own right, needing careful maintenance, and ideally it should be part of a crop rotation. That high ground tends to be covered in permanent grassland results from the inability to cultivate other crops, and as such is less 'efficient' than lowland grass in supporting grazing animals. This means that more sheep per acre can be reared in the flat French country than in hilly Wales, so Welsh farmers are at an inevitable disadvantage compared to the French, other things being equal. Regrettable, but true! Neither can the Rocky Mountains compete with the verdant pastures of New England.

Regular maintenance of grazing land includes the removal of nettles, thistles, trees and brambles and the upkeep of perimeter boundaries such as walls, hedges or fences.

Most grass ceases growth between November and March in the northern hemisphere, and whilst sheep can continue to graze during the winter, the density of animals at this time must be kept low. Other ruminant mammals (these have the ability to convert insoluble and usually indigestible grass cellulose to useful nutrients) include deer, goats and cattle, that often require artificial feed in winter. Traditionally this feed has been mainly silage produced by storing the previous summer's cut grass crop.

A well-kept grassland, ideally surrounded by hedges or woodland, should provide good long-term grazing and the permanent cover of vegetation should 'fix' carbon dioxide from the air, and so help counter the greenhouse effect. This effect is due to gases in the atmosphere that trap heat and so raise the temperature of the Earth. The fertility of the soil should be maintained by the grazing animals' excreta, enhanced by wildlife whose sanctuary is provided by boundary trees and hedges. Rabbits, hares, mice, moles, hedgehogs, badgers and foxes should all flourish in such an environment, together with a huge variety of wild plants. Insects will flourish too, with butterflies the most obvious.

One of the most spectacular types of this grassland is the Alpine meadow, between about 3,000 and 8,000 feet above sea level. The grass is cut just once, in the summer after the riot of alpine blooms. At lower altitudes grass can be cropped as many as three times a year. But, to maintain such a yield, artificial fertilisers may be needed.

This approach to grass management should provide safe and nutritionally excellent food for mammals and some birds; it should provide positive environmental benefit, a satisfying livelihood for the farmer, and a source of pleasure for the itinerant city dweller.

In previous decades, such grassland would have been used for rearing chickens, geese, ducks and pigs, in addition to ruminants. But now it is under the threat of extinction. This is mainly because of the move to intensive rearing of food mammals and birds in sheds, using artificial feed produced from cereals. This results in a cheaper final product compared with free-range husbandry. We seem to have (or used to have) a limitless capacity for wanting cheap meat. We shall be returning to this issue, particularly in Chapter 17.

Formation of deserts

The other problem concerns the overgrazing of vegetation, often associated with periods of near or actual drought.

Consider a territory that is situated between the tropics and the cooler temperate zones, and with low to moderate rainfall. The natural, fully developed vegetation (known as climax) will tend to consist mainly of narrow and broadleaved trees and shrubs. The leaves will release a considerable amount of water vapour into the air, which in turn may produce cloud and rain.

Over the years an ecological balance will be reached with the tree roots able to penetrate to the ground moisture, the upper boundary of which is referred to as the water table.

Suppose these trees are felled and grazing established. Initially the grass may grow, and there may be adequate moisture left in the soil to support, say, cattle. But inevitably, from time to time, phases of drier than usual climate will occur. Then the grass will not grow, the cattle will continue to graze, the vegetation will disappear. Because of the reduced amount of water vapour released into the air by the disappearing vegetation, the rainfall will lessen further, the water table will drop and the beginnings of a desert become evident. Even trees newly planted may not survive because their roots cannot reach the water table.

This sequence of events is not hypothetical. It has actually happened in Central and Eastern Africa and in the USA, notably in Iowa.

The message from this sequence of events is very clear. We should treat grassland as a very precious crop. Great care should be given to decide whether it should be created in the first place, and constant cultivation and management provided subsequently.

Cereals; wheat

Cereals are specially adapted grasses. Wild barley and oats abound in the temperate world and their seed heads look similar to the cultivated varieties. We eat the seeds of the cereal crop, and being seeds they must have a good range of nutrients to initiate new plants. It is only when a plant is, say, 2–6 inches high, that it will have exhausted the nutrients in the seed.

Cereals are the most important single type of crop for the entire human population, and of all those grown, just three – wheat, rice and maize – make up by weight 80% of the cereals that we eat. The remainder include barley, oats and rye. Wheat is grown most extensively in North America and Europe, and has been very thoroughly researched and developed to produce maximum yields.

My first memory of wheat goes back to the late 1940s, when staying with my grandparents in a cottage in the East of England. At that time the seed and stalks of wheat were cut and put into stooks on the ground to dry, before being loaded into a truck for taking to the plant to extract the grains. This was an inefficient process, quite a few ears being dropped on the ground.

So we went 'gleaning', which means rather furtively collecting the dropped wheat to feed to the chickens scratching around in the orchard. I still do not know if this activity was approved by the farmer, but my grandparents' rules were clear: on no account glean any of the wheat in the stooks.

Of course the cutting of the ripe wheat and extraction of the seed are performed now in one operation by the combine harvester and this accounts for one of the many changes seen in the type of wheat grown in the last 40 years – shortening of the stalk, desirable because of increased resistance to wind damage. Whilst modern harvesters can process wheat lying on the ground, there is a reduction of yield, due to damp causing fungal infection and also sprouting of the seed before it is harvested.

The other changes in wheat husbandry have all been aimed at improving efficiency. The goal seems to have been to increase maximum weight per acre of crop that matures within the shortest time possible. Of secondary importance are resistance to disease and storing quality. Even less consideration has been given to changes in the nutritional content and to the environmental impact of the new varieties.

The striving towards greater efficiency has been achieved by an enormous amount of research in the following areas. First, the search for high yielding (more and larger grains per seed head), early ripening, and optimum timing and density of planting. Because wheat takes a great deal of nourishment out of the soil, and returns little, the idea has been popular for years of using genetic engineering to creating strains whose roots could convert nitrogen gas in the air to nitrogen salts in the soil, that could then be used by the plant. This nitrogen fixing is achieved by beans, peas and clover through specialised bacteria growing around their roots. Indeed, one established cropping method is to plant nitrogen-fixing clover alongside rows of wheat, but so far the creation of a type of hybrid – part clover, part wheat – still seems some way off commercial use.

Even if wheat was developed that could fix nitrogen – and so reduce the need for nitrogenous fertiliser – other fertilising chemicals would still be needed. In particular, the supply of phosphates will pose a problem because there is an impending world shortage of this material in a form that is easily mined or extracted.

The modern types of wheat are indeed high yielding, they can succeed in relatively cold damp climates, and they do give a prolific yield and ripen early. They provide a secure, high volume, profitable product for the farmer. Wheat is easy to store and is probably the most important international food commodity, with trading (and speculation) in all the major financial markets. The research and development have paid off.

So what are the problems? The first seems to be caused by the surplus. We, the consumers, at least many of us, seem to want to eat only part of the wheat seed: the white inside (the endosperm), not the darker outside layers rich in fibre, vitamins and minerals. One reason for our preference for white over brown bread is that in general white bread can be lighter in weight than brown, because the yeast acts on more of the flour.

True, there are a substantial number of consumers who understand and prefer wholemeal bread, but surely to make white bread requires unnecessary cost and energy (to remove the outside coatings of the grain)? One of the problems of freedom of choice associated with an abundance of food is that we can select what is not best for us!

Successful high-yield production of wheat and other cereal crops requires fertilisers, and control of diseases and parasite attacks. Stored wheat, particularly under hot, humid conditions, can generate further parasitic problems, with the need for further chemicals. Certainly some chemicals have been shown to be dangerous and are no longer used, for example DDT. Others, we hope, are safe, but there is no real way of establishing this with certainty. But the approach most countries currently use towards growing wheat must damage the structure and quality of the soil over the years, particularly regarding the amount of organic matter that can be so important for water retention.

There are also potential problems as a result of early maturation of modern varieties. In a good summer, the crop can be harvested in July, with vast expanses of ground left bare, or nearly so, for 8–9 months. During this time, the soil nutrients will be dissolved by rainwater and lost from the ground, and the absence of vegetation will aggravate the greenhouse effect because there will be no extraction from the atmosphere of carbon dioxide that is constantly increasing due to burning fossil fuels. Moreover, without vegetation, the dry surface soil can literally be blown into the air.

Other cereals

Maize, or corn, is grown widely in warmish, temperate climates such as in the USA and central Europe, for both human and animal feed. Many different varieties exist, some crops producing a succession of cobs in late summer, others just one. The plant is grown in small areas, in blocks rather than rows to aid wind pollination; and it is also a greedy plant, using up much moisture and nutrients from the ground. However, the yield can be very high.

Nutritionally, corn has one problem in that vitamin B_3 (niacin) is not absorbed by the human body after eating it, and if corn is the only source of this vitamin, a disease, pellagra, can occur. However, this is rare as several other foods such as meat, fish, dairy products and potatoes all contain niacin.

Some varieties of maize produce highly coloured kernels; blue, violet or red and are grown only for decoration.

Cornflakes are made by combining the dried, powdered maize with malt extract, minerals, vitamins and water into a thick slurry, after which it is roasted.

Powdered corn or meal can also be used to make bread, more popular in the USA than the UK.

Popcorn is made from a particular variety with a thick and tough outer seed coating. As a result of pressure generated inside the kernel by heat, a critical point occurs when the coating is breached, with a minor explosion, resulting in the inside endosperm rapidly expanding as the corn 'pops'.

Maize is therefore potentially a prolific crop, relatively free from disease, but very demanding for fertilisers.

Rice is also an international commodity, with most of the world crop grown in the paddy fields of the East. It could be grown more widely, but its need for intensive labour, particularly at planting, can make its cultivation unattractive to the developed world. However, a considerable amount of rice is grown in the southern USA and is harvested 'mechanically'; some of this finds its way to Europe. The produce containing the whole rice seed is recommended and is often referred to as 'brown rice'.

Nutritionally, rice is excellent, but as with bread, the preference for the white grain type denuded of its coating, rather than the brown, is puzzling.

Wild rice is increasingly fashionable, particularly among conservationists! Whilst being a grass, it is not a true rice. Its origins go back to the American Indians around the Great Lakes, where it is still grown and harvested manually. It is obviously expensive to the consumer, but very positive in flavour and colour; the blackish grains turn purplish on cooking. Wild rice is frequently added to 'ordinary' rice to give more appeal. Recommended.

Barley has been subjected much less to improvement or selection compared with wheat, and its use is much more limited. The harvested whole grains tend to be tough and need thorough soaking before use. Whole barley is mainly required in the first stages of brewing beer. Pearl barley is the dried inside of the seed, and gives a subtle flavour and also a soft silvery texture to soups.

Barley flour can be used in bread, but the dough tends not to rise unless a substantial amount of wheat flour is also present.

Rye had virtually disappeared from many countries earlier this century, the crop being less productive than other cereals. Rye bread is usually made from part wheat flour and part whole rye flour and is popular in Eastern Europe. Nutritionally it is excellent and would seem set for a comeback. This is already occurring in the USA.

Oats are used mainly in cookie, biscuit and cereal manufacture and impart a crumbly, chewy texture to the finished product. The relatively high fibre and protein content of whole oats has suggested desirable nutritional qualities, although feelings about other components, such as sugar and salt in cookies and biscuits, might not be so positive.

There are other cereals: bulgar wheat has its enthusiasts, and no doubt other species will be farmed. They can all be expected to provide good basic nutrition, but will be greedy for soil nutrients.

Potatoes

The three main root crops grown in the world are potatoes, sweet potatoes and cassava.

Potatoes first arrived in Europe from the mountains of South America in Shakespeare's day, and subsequently were distributed to many other countries, including the New World. Potatoes are fairly tolerant of soil and climate, but yield prolifically only on highly fertile ground. New potatoes are surely one of the delights of eating! When freshly dug, the skins just rub off.

The purpose of scraping or scrubbing is to remove just the very outside covering, leaving behind the yellowish and dense layer just underneath. It is here that much of the flavour of the potato is found.

But as new potatoes are stored, the outside layer becomes adherent to the inside so proper peeling is required to remove it. By the 1970s we were beginning to leave this on before cooking. Then rumours that potato skins might cause diverse health problems abounded. So it was back to peeling! These rumours have now abated, so it's potato skins once more! Restaurants commonly serve crisped potato skins as a delicacy! Many people do find the skin of the well-baked potato definitely enjoyable, although, if it gets really burnt it will become almost charcoal and tastes a bit metallic.

The appeal of the genuine oven-baked potatoes is the smell of the half-burnt skins – even if you do not eat them. The microwaved version, whilst still described as a jacket or even oven potato cannot be considered 'baked' and the skin never has the aroma or crispness of the real product. If the menu in a restaurant describes a potato as 'oven baked', and I am served with a tasteless microwaved lump, I send it straight back.

Of course restaurants do have problems in cooking a perfect jacket potato, that might take an hour and a half for a customer who wants immediate service. They can try one of several dodges. One is to oven-bake a batch of potatoes in advance, then keep them refrigerated. When requested, the potato receives a brief massacre in the microwave before serving. Or the potatoes can be microwaved first and then briefly grilled. This can be quite reasonable. But ideally, potatoes should be oven-baked in advance to supply the expected demand. Incidentally, the reason that potatoes are rapidly cooked in a microwave 'oven' is that their high water and low salt content encourage penetration of the microwaves.

Potatoes have increasingly been the subject of food processing – for example, to make thin, fried potato slices, known as crisps in the UK and chips in the USA. Indeed, whole crops are purchased in advance by the food processing industry. Farmers therefore grow varieties suitable for these purposes, which may well not be suitable for baking or for eating as 'new' potatoes. This explains some of the difficulties consumers may experience in actually finding particular types of potato.

It is gratifying to see a return to some of the older varieties selected over the decades for primary cooking, rather than manufacture into microwavable chips! The reason for placing so much emphasis on the optimum cooking of potatoes is that when care is taken, they must be one of the ideal foods. They are currently being ruined in many types of crisp or chip and the farmer is persuaded by the processors to produce inferior types.

The important principle seems to have been established: the way we process and cook any food determines whether it is desirable and whether crops are grown. We should cook potatoes better, and then we will want to grow more for uses other than for crisps or chips.

Nuts

Nuts are one of the food industry's successes. Do you remember the days of famine after the Second World War, when nuts were available only for Christmas? I can recall the trouble we had getting them out of their shells. The hazelnut was too small for the nutcrackers and needed to be attacked by a hammer. More often than not it went ricocheting around the room, to end up out of reach under a chest; sometimes the hammer blow was so successful that it reduced the nut to pulp. Brazil nuts usually broke the crackers or someone's teeth and the only time a walnut was delivered intact from its shell, it was mouldy. Almonds were easy to extract from their shells but were usually too shrivelled and bitter to eat.

You can now buy selections of unshelled nuts from some retailers, but the variety of individual packeted nuts available is remarkable: peanuts, cashews, walnuts, almonds, brazils, hazelnuts, macadamia, pecans, pistachios and pinenuts, to name a few. Macadamia nuts are large enough to slice or grate and give salads and fruits some sparkle.

From the nutritional point of view, nuts have everything: protein, minerals, polyunsaturated fat, fibre and energy. Don't try this experiment, but I guess you could live on nuts alone. Because of their low water content, nuts are pretty safe unless something stupid is done with them, as occurred when inadequately heated canned hazelnut purée was added to yogurt and caused botulism in the UK in 1989.

So the variety of available nuts is now impressive. But are not some being ruined with too many flavourings, mixtures and sales gimmicks? Dry roast peanuts paved the way for honey, chilli, Bombay (I wonder what they would make of these in India) and yogurt flavours. The nuts were also put inside coatings and ever more and more types of chocolate. Their marketing seems to be going the same way as crisps (chips). It is now possible to buy 'salt and vinegar flavour dry roast peanuts'. Some nuts seem to be getting too salty.

Many nuts are obtained from Third World countries, using

the available cheap labour, and then transported through a long and tortuous route.

Sugar

Sugar cane and, more specifically, sugar beet are probably the two most unnecessary and wasteful crops grown. Sugar, apart from supplying energy, has no essential role in our nutrition, and is only needed in such quantity as to make so much of our processed food appealing. The sugar industry points out that it is natural and safe. It also claims that sugars are essential components of every cell, so we need sugar in the diet. Not true. The human body can quite easily make sugar from other substances in food, such as starch. Why not eat starch as in potatoes or cereals, that need no extraction or processing?

Sugar cane needs wet and hot growing conditions followed by a reasonably dry ripening climate. These conditions are most often met in the Third World, which accounts for the cheapness of cane sugar.

Sugar beet is grown in temperate climates and typically 100 g of sugar is present in 700 g of the whole root. Sugar beet can, however, flourish on moderately impoverished soils and can, like potatoes, provide a useful crop to rotate with cereals.

Fruit and vegetables

In developed countries, most fruits and vegetables are available continuously, even if quite out of season for the locality. I can see the attraction of this, but it leads to some problems. Does not our taste become jaded for a regularly eaten product? Do we not enjoy something all the more if we have had a long wait for it? Are we not spoiled by continual abundance? Is the quality of the out-of-season product always acceptable? Are we not over-influenced by appearance and packaging, rather than taste, smell and nutritional content?

Consider tomatoes. There is something seriously wrong with some of our tomatoes. True, there is more choice, at least in size, that can vary from that of a pea to a misshapen grapefruit. Often just the tiny ones seem to have a positive tomato smell. But what of the flavour and the juiciness and the smell of the others? The skins are often tough, the flesh is woolly, soft or limp and the juices can be replaced by air. Do we really want more, different types of a dull, tasteless fruit? Would you recognise a tomato were it not for its colour or thick skin? Whose fault is it, and why has it happened? The heart of the problem, I think, is that the cheapness of the food is all important to the consumer. Cheap tomatoes can look like expensive tomatoes, and if they are enclosed in a plastic box, there will be little smell anyway. The cultivation can involve highly unnatural growing conditions, usually with water containing artificial fertiliser (known as hydroponics). The whole system is geared to a product that grows as rapidly as possible and then survives its prolonged journey to the retailers and restaurants. It is no wonder that almost all the available tomatoes are under-ripe!

So why do we keep buying so many of these tasteless products? The answer, I suspect, is because of their colour, that enlivens many a drab plate. In particular, tomatoes are *de rigueur* for garnishes for sandwiches, etc. A garnish seems to be something that you put next to an unappetising food item but which you don't actually eat!

Would it not be preferable to buy better quality local products when in season and actually enjoy eating them? Even if they were slightly more expensive, it would be worth it because there would be less waste.

Much of what concerns tomatoes also applies to other salad items such as cucumbers, lettuces and radishes purchased from supermarkets. It is no wonder that more and more supermarkets are disguising poor salads by mixing them into coleslaw-type products!

Mushrooms

Mushrooms seem to be as popular as ever. But is this justified? What about their taste? We have them all the year round now. How?

In the wild, the mushroom is the fruiting extravagance of a thread-like white soil fungus that produces a complex invisible network underground. In late summer and early autumn, when rain moistens the still-warm soil in meadows and sparse woodland, the mushroom appears first as a tiny button, and then the stalk pushes upwards with its canopy developing, with millions of spores on the underneath black-brown gills. These spores are carried on the wind to start new fungi. The flavour of the natural mushroom is found mainly in those dark parts.

I can recall about 40 years ago tramping the fields in the early autumn looking for wild mushrooms. Dark green grass fertilised by animals suggested where to look. We usually went out very early in the morning, so our finds could be converted to breakfast. Mushrooms on toast then were a real luxury.

Wild mushrooms grow best in ground enriched with rotting vegetation or animal manure, and they can form without any light, as they contain none of the green pigment chlorophyll. So it is relatively easy to grow mushrooms artificially at any time of the year. All that is needed is a good supply of rotting debris such as horse manure, moisture, and heat provided within sheds. The commercial growers have to take precautions against contaminating fungi from taking over (incidentally, these might taste better than the mushrooms).

Because of the risk of damage to the fully developed mushroom during its transport, most artificial mushrooms were initially harvested when still immature, or button-shaped. In the supermarkets and greengrocers, these still predominate.

So far, so good: a seemingly sound operation based purely on business criteria. But what of the mushroom's taste? Perhaps the renewed interest in, and availability of, large flat open mushrooms indicates that there is some concern about the lack of flavour of the button mushrooms. Chinese restaurants seem to carry these to extremes, with 'straw' mushrooms being as little as 10 mm in diameter. But on the positive side, it must be the lack of taste of cultivated mushrooms that has recently stimulated the introduction of further types of edible fungi; not that these ever needed to be re-introduced in France!

Other crops

Space prevents detailed comment on other crops. But peas and beans (pulses), and the seeds yielding vegetable oil, such as sunflower, are all important crops and will be considered in detail in Chapter 15 with regard to their nutritional value.

Baked beans said to be one of the British Prime Minister's favourite meals, do need a mention. North Americans may be puzzled by the popularity of this genre. They do possess the overriding criterion for British appeal: they are cheap. But what exactly are baked beans? Generally, they are derived from the seeds of what is essentially a French bean, the pods having been allowed to ripen for longer than when eaten as a vegetable. The bean seeds are dried, are now known as haricot beans, and join the international bulk food trade. Haricot beans are grown in many countries, particularly in the Third World. When the haricot beans are rehydrated, sauced and seasoned, heated and canned, they are known as baked beans, although why they should be described simply as 'baked' is something of a mystery.

Beans are rich in protein, carbohydrate, unsaturated fats, fibre and some vitamins. Some salt and sugar are added, and if there is a question mark over their nutrients, it is the amount of added salt and sugar.

Oilseed rape

Some years ago a visitor from London was with us looking at
one of many fields of sugar beet in Norfolk, UK. He asked, 'Why
do they grow so much spinach in Norfolk?' I suppose the foliage
of beet and spinach are quite similar! Next year he did even
better. Surveying a field of oilseed rape in flower, he asked why
they grew so much mustard in Norfolk!

Take a sunny May day in almost any part of lowland Britain
and you will be dazzled by the yellow of oilseed rape; England's
green and pleasant land at this season has become England's
yellow and (hopefully) still pleasant land. Despite warnings, many
farms persist in growing oilseed rape in successive years on the
same land. This increases the risk of attack by nematode worms.
There is no reason why this crop cannot be part of a rotation.
However, one problem has been its toxicity to wild deer. Early
worries about the inevitability of many diseases affecting oilseed
rape seem so far unfounded. The crop has the advantage of
nitrogen fixation capability.

The oil is extracted from the seeds and is high in polyunsaturates and mono-unsaturates; it is known as canola oil and is used
in intensive livestock feeding. However, in the USA, canola is
increasingly used as an alternative to sunflower or cottonseed for
cooking, dressings and so on.

Organic farming

The realisation that intensive crop production as practised in
the USA and much of Europe would, sooner or later, become
non-sustainable, is the basis for organic farming. It has also been
described as biological farming, regenerative and sustainable agriculture. Most of the public see organic farming as the production
of food without artificial fertilisers or insecticides and other
chemicals, but the real point in this approach is to provide a

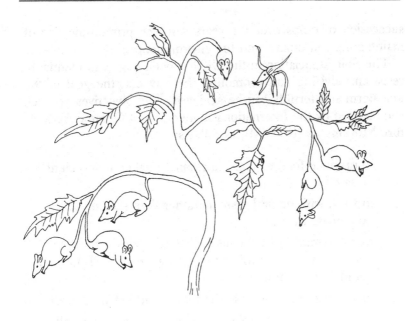

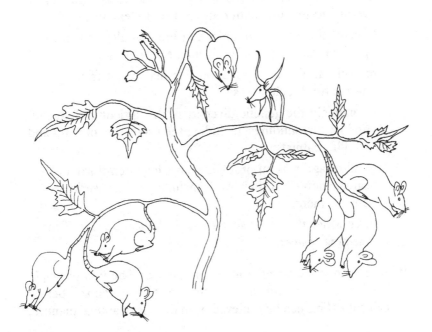

succession of crops over the years with the prime objective of maintaining soil quality and the environment.

The Soil Association in the UK has been hugely influential in research, advising and promoting the way that the goal of the long-term soil fertility can be achieved. The objectives are well summarised by the International Federation of Organic Agriculture Movements (IFOAM), as follows:

  to produce food of high nutritional quality in sufficient quantity;

  to work with natural systems rather than seeking to dominate them;

  to encourage and enhance biological cycles within the farming system, involving microorganisms, soil flora and fauna, plants and animals;

  to maintain and increase the long-term fertility of soils;

  to use as far as possible renewable resources in locally organised agricultural systems;

  to work as much as possible within a closed system with regard to organic matter and nutrient elements;

  to give all livestock conditions of life that allow them to perform all aspects of their innate behaviour;

  to avoid all forms of pollution that may result from agricultural techniques;

  to maintain the genetic diversity of the agricultural system and its surroundings, including the protection of plant and wildlife habitats;

  to allow agricultural producers an adequate return and satisfaction from their work including a safe working environment;

  to consider the wider social and ecological impact of the farming system.

What does organic farming mean in practice? The first principle is that farmers must put the long-term fertility of the soil before quick profit. This can be achieved from the use of animal manure, as with the mixed arable/livestock farms. Consider, for example, a lowland field. Let us begin with it as grassland and cattle grazing according to the rate of grass growth. Next it could be used for winter wheat, sown in the previous autumn. The wheat would be harvested in July, and by August a rapid 'catch crop' of turnips could be sown (if they failed to mature, they could be ploughed in as green manure).

Potatoes or barley could be planted in the spring of the third year, with the land sown in the early autumn with grass for cattle grazing once more from the early summer of the fourth year. Alternatively, the ground could be seeded with grass/clover and left fallow for the entire year.

The value of crop rotation is well known and should mean that little or no artificial fertiliser was required. Also, for almost all of the time, the ground would be occupied by vegetation. This is important, as the roots help hold water and nutrients which otherwise tend to be carried away, dissolved in rainwater, and prevent loss of the soil through wind erosion. The need for the use of chemical sprays against diseases and pests should be lessened by the crop rotation which would make any build-up in disease agents unlikely (most crops have their own, and often only their own, set of problems).

Pests would be further discouraged by sowing the crop seed sufficiently thin to encourage healthy plants to resist the disease agents.

With a system such as this – and of course details must vary with locality – the fertility of the soil should be safeguarded more or less permanently. Only occasionally may some chemicals be needed – for example, lime to counter excess acidity.

Consider the present typical intensive lowland wheat farm. There will be no grazing livestock. The farmer will have a contract to sell so much wheat annually and is rewarded by guarantees of sale. Each year the soil will be drilled (sown) in autumn or early spring and harvested in July, with several months of exposed soil. Previously the unwanted straw might have been burnt. Now it tends to accumulate in ugly heaps. Apparently it is simply not worth the effort of turning it into compost and returning its nutrients to the soil. The seed will be sown as closely as is needed to give maximum yield, and fertiliser applied regularly. Any evidence of infestation will be treated by chemical sprays.

So what are the outcomes of the organic and intensive farms? The first difference is that the profit from the intensive farm is greater than that from the organic farm, because the long-term condition of the soil does not fit comfortably into the financial calculations.

Then organic produce is more expensive and might not actually look as perfect as the intensive. Most consumers do not feel a responsibility to the future state of the ground when they purchase some microwave chips (fries)!

This explains why organic produce has really failed to make substantial inroads into the retail sector. Indeed, one major UK food retailer abandoned organic produce altogether in 1991. Unfortunately, organic produce may not taste any better than the chemical, and has not been proved to be better for you.

But the one benefit attributed to organic produce over intensive is the freedom from chemical sprays. This advantage remains somewhat speculative, because it is not possible to perform an experiment to find out for certain whether the repetitive consumption of small amounts of some of these chemicals is going to cause ill health. For the ruling on the safety of most chemicals in use today, the 'jury' is still out.

But the 'jury' has returned its verdict on the effect of intensive farming on soil and the environment. The farmers, politicians and the consumers who also support the system have been found guilty of environmental vandalism.

Genetic engineering:
how to create a blue orange

Not content with improving yields and sizes by natural selection and cross-breeding, we have learnt to create totally new life forms by genetic engineering. This technique dates back to the mid-1970s and was very much an American invention. The following section is rather technical. Apologies, but maybe this is why the genetic engineers have pulled the wool over our eyes for so long!

Genetic engineering enables us to create new forms of animal and plant life, and products from these can certainly be useful, such as human insulin. But other products may be unnecessary, irrelevant or even dangerous.

Let us look at the hypothetical (I think?) instance of a company planning to create and sell sky-blue coloured oranges. How would it proceed and what problems might be encountered?

The idea that consumers might be attracted to blue oranges is not completely crazy. After all, we have striven with considerable success to achieve black tulips, yellow sweet peas, pink daffodils, blue primroses, and blood-red oranges. White chocolate seems in fashion, so why not blue oranges?

The company would have to plan a major investment into the project over several years. It would probably need first to rid the orange of its natural colour and then somehow insert a group of genes into the plant that enabled it to synthesise blue colour.

The first step would be to identify an 'albino' orange, or a plant that produced only tiny amounts of natural pigment. These may occur as natural 'sports' from time to time as a result of chance loss of the region of the DNA (the genetic material) of the plant responsible for the colour. This can happen through ultraviolet light from the sun scoring a direct hit on the specific segment of DNA in, say, a pollen grain.

But 'albino' oranges might not occur in nature, in which case they can be created as a result of treating isolated cells of the orange plant with chemicals to knock out or delete genes (regions of DNA). In the later part of the experiments these individual cells will probably be needed anyway so the company would at the outset establish the technology.

Most people are familiar with the methods of growing bacteria on the surface of nutrient gels in glass or plastic dishes (it was one of those plates, contaminated with a fungus, that produced penicillin and so started the antibiotic era). The clumps of bacteria are visible as blobs after growing for days, although each individual bacterium is visible only under the microscope.

Cells of plants can also be grown by this method. These may be obtained from growing roots, ground up and dispersed into

fluids and suspended on the surface of materials akin to moist
blotting paper. If appropriate nutrients, moisture and the correct
amount of light and heat are supplied, the individual plant cells,
although invisible, start to grow. The major problem, contamination by bacteria and fungi, is countered by meticulous hygiene
and the addition of chemicals to the nutrient fluid that prevents
the growth of these organisms, but does not damage the tiny
plant.

Amazingly, a single cell from a plant root can ultimately yield
an entire plant with stem, leaves, flowers and fruit. This is, perhaps, not so amazing when it is appreciated that the fertilised
human egg also starts as a single cell that then develops into a
person.

The process whereby a single cell gives rise to other cells of
diverse properties is known as differentiation, and the first
changes in our tiny orange tree will be the upward growing cells
that will become a shoot and the downward growth, the root.
The tiny embryo plant is exceedingly delicate at this stage but it
gradually acquires vigour and can be transplanted from its chemical nutrients to soil in the laboratory and ultimately to its final
habitat either in glasshouses or outside. This technique requires
skill, is costly and is slow, and there is always the danger of
contamination. It does, however, enable a single plant cell (or
animal cell, for that matter) to be altered, and the altered DNA
to become permanently part of the plant's make-up and then be
inherited, through the seeds, by subsequent generations.

To produce an 'albino' orange, these plant cells are exposed
to chemicals (or radiation such as X-rays) known to damage discrete parts or focal areas of DNA, and it is hoped that on grounds
of probability, some cells will experience damage just to the part
of their DNA responsible for colour, so the cells will still be able
to grow. Unfortunately this may not occur often, and it means
that thousands or millions of cells will have to be grown on to
tiny embryo plants, to seedlings, and then on to mature plants,
when, years later, the albino orange will be spotted. A very expensive process.

Let us suppose this experiment is indeed successful and cells
from the root of the 'albino' orange are now available, and can

be separated, dispersed, and then grown on. The next step is to insert into one of these cells the genetic information that will enable it to synthesise blue pigment. For this a number of genes will have to be found that already carry out this function in another plant, and it will be hoped that they will also function (that means cause the appropriate cells of the plant to make the blue colour) in the 'albino' orange.

The company will probably try several true blue flowering plants: perhaps forget-me-not, anchusa or delphinium. Cells from these plants will be grown in artificial culture as already described, and all the DNA from such a culture collected and purified. This is a straightforward step because DNA is easy to separate from proteins and, being relatively dense, it can be pelleted to the bottom of a tube by a centrifuge. This DNA contains all the genes in the plants, including the relatively small part that makes the blue colour.

The problem is that it will probably be impossible to identify and concentrate just the 'blue' DNA. Instead, enzymes have to be used to break all the DNA into small pieces in the hope that some of the resulting fragments will contain all the blue genes but not too much else.

Most cells can actually take up these pieces of DNA, so when the 'albino' orange cells are showered with DNA fragments, some, including the 'blue' genes, are likely to enter the 'albino' cells and become attached and then permanently incorporated into its DNA.

The problems are not over, because it is not possible at this stage to know which tiny developing plant embryo will produce blue fruit. Another lengthy process begins, requiring that once more thousands or millions of orange trees are grown to see if any produce blue fruit.

The cost of this exercise is therefore very substantial and it is no wonder that genetic engineering companies attempt to patent their products in order to recoup their high costs – with interest. Without such patents, much of the profits, and indeed the stimulus for genetic engineering, would disappear, because once a new life form was created and identified it could then be used for propagation very rapidly.

What are the problems and the dangers? The main concern is that the whole process is generally hit-and-miss: other substances in the orange tree may be altered inadvertently and unknowingly. For example, in the creation of the 'albino' orange, the chemical may also damage other regions of the DNA and new chemicals be produced in the orange fruit. Or, when the fragments of blue DNA are introduced, the function of the existing DNA could be adversely affected in all sorts of ways, to produce dangerous substances. Whenever a small piece of DNA is inserted into a longer one, there is a risk that the regions just next to the place of the insertion are made non-functional, which means that a part of the normal protein or sugar made in the fruit could be missed and the resultant substance could be poisonous to us.

Our experience with eating oranges over the years has established them as safe: in contrast we have learnt that nightshade, aconite and some fungi are dangerous. Foxgloves produce a substance called digitalis which, under very carefully controlled conditions, is of value in treating some heart conditions. But for most people, certainly if eaten regularly, it is unquestionably poisonous.

How on earth can the safety of a blue orange be established in practice? The simple answer is that it cannot, and herein lies the whole dilemma with much of genetic engineering.

Novel products which do not produce any obvious benefit must always be beset with safety fears, and we must proceed with extreme caution with genetic manipulation of our food. Genetically engineered tomatoes the size of water-melons may not be good for us, but we cannot and will not know for years whether this is so.

Unless we really do need a product created through genetic engineering technology, then there should be no persuance of that substance. Do we *need* blue oranges or giant tomatoes?

But to end this section on an optimistic note, where we do need such a product, the theoretical risks or dangers may well be much less important than the proven benefits.

Genetically engineered growth hormone is a valuable drug and has now, for very good reasons, replaced the natural hormone.

This used to be obtained from the pituitary gland, in the base of the brain, from people after death; it became contaminated with the infectious agent causing Creutzfeldt–Jakob disease (see Chapter 9) and has already infected fatally at least 20 people. The growth hormone produced by genetic engineering poses no such risk.

In conclusion, we do have the facility to grow a wealth of plant crops, some having been improved by hybridisation and selection. But we frequently grow the wrong crops and rarely consider now the effects of cultivation on long-term soil fertility. Genetically engineered crops are fraught with hazards.

I award the human race 6 out of 10 for crops. We could do better.

Recommended cereals, vegetables and fruit and their products

Wheat (whole)
Oats
Rice (brown)
Potatoes (most)
Maize (corn)
Rye
Nuts
Seasonal vegetables
Italian tomatoes
Beans (and the other pulses)
Local fruit in season
Most citrus fruit
Most imported tropical fruit
Field mushrooms and the edible fungi
Vegetable oils: sunflower, corn, olive, canola, cottonseed, groundnut, soya, sesame, safflower

Some crops and products difficult to recommend

Sugar
Some apples
Palm or coconut oil
Fried potatoes
Giant tomatoes
Button mushrooms
Out-of-season vegetables

2
Early meat

Introduction

Our ancestors presumably added meat to their basic diet because of a shortage of 'vegetables' and fruit. It is worth pointing out here that we can quite happily live on a diet without meat at all (Chapter 8), but I do not think we would flourish on a diet composed solely of meat (although we might on fish).

In this chapter we discuss essentially the origins of meat production. In the next two chapters, the move to intensive rearing is discussed.

Ethnic attitudes towards animals

Many mammals, birds and fish have been reared for so long under artificial conditions that we may have forgotten their origins. Indeed, animals such as pigs are now so often housed in sheds that we may even come to think that this is their natural domain. When did you last see a live turkey? The situation is confused by the rapid changes produced by breeding, so that in many ways animals today may not resemble their ancestors. Genetic engineering can be used to accelerate the progress, and can create terrible suffering in the new forms produced.

Each type of human society, for various and often understood reasons, tends to develop particular attitudes to the varying species of mammals it meets. Thus, in the UK meat is eaten mainly from just three mammals: cattle, pigs and sheep. Chickens, turkeys and ducks provide most of the poultry, and among deep-sea fish, white-fleshed species are preferred to darker-coloured. In the USA, it is beef, then more beef, and then even more beef.

However, goats are extensively reared for food in Africa, and many religious groups do not eat certain species: for example, the Jewish and Muslim people do not eat pig products and the Hindus do not eat cattle.

Moreover, there are differences amongst countries, races or even tribes over the acceptability or desirability of other sources of meat. For example, ostrich farming is developing in some countries, such as South Africa; the Chinese eat dogs; the French eat frogs and snails. The Italians delight in thrushes, the British in pheasants and grouse, and so on. Why these differences have occurred is not at all clear, although the Jewish aversion for pork has a biblical basis.

We seem to express three different types of attitude to animals and birds. One is that of reverence, as is shown to our companion animals, such as dogs and cats. Then there are the more neutral feelings of indifference, tolerance, curiosity or, indeed, affection towards non-domesticated animals such as rabbits, many wild birds, and hedgehogs. The third category concerns those we eat, where we try to dissociate our eating needs from the circumstances of the animals' rearing and slaughter. These social attitudes are deep-seated, but generally do not have a logical basis. Some people tend to adopt the eating habits of their country of residence, regardless of their own origins.

But there is no dietary reason why we should not keep chickens as pets and eat cats. The latter suggestion may be revolting because of our social attitude, not because cat meat would be bad for us.

Whether we should eat so much meat is unquestionably a matter of considerable current concern and the amount eaten will probably decrease in the future. This is because not only have our food animals been selected somewhat arbitrarily, but also there is increasing evidence that meat is in several ways inferior to a vegetable diet (see Chapters 8 and 17).

Ruminants

The ancestors of the ruminants – cattle, sheep, goats and deer – would have inhabited sparse woodland or grass plains, notably in Africa. The last 30 years have seen the increasing control of deer. These animals have not been subjected to major breeding or to selection – at least, not yet. In the wild, deer inhabit woodland for shelter, but certainly venture cautiously into open spaces for grazing. They are sociable animals and their large groups provide protection against predators. We are now rearing deer in controlled spaces, mainly open fields supported by sheds, and are fed additional feed supplements, mainly in the winter, to accelerate weight gain.

Sheep have since prehistoric times grazed on open grassland, particularly at high altitudes. Hill walkers will be familiar with the hard droppings of sheep – hard because the animal strives to retain every drop of water it can. Indeed, the sheep is wonderfully adapted to water conservation, getting most of its daily intake from the moisture within the grass itself, from early morning dew or from rain.

The main alterations man has imposed on sheep are in the colour of the wool, now less important because of the decline in the wool industry, and in the numbers of lambs born to each ewe. In nature, the female sheep (ewes), like human beings, produce predominantly single offspring. Now twins, and occasionally triplets, are the order of the day. However, sheep have proved extremely stubborn in other ways, resisting the move to intensive rearing in sheds. If this is attempted, they just seem to fade away and die, often for no apparent reason. Another problem has been the timing of birth of the lambs, typically February–March, not ideal in the Northern Hemisphere because of shortage of grass and the risk of heavy snow.

It is in these areas that the hybridisers and genetic engineers are currently giving most attention. It remains to be seen whether an animal – say, producing triplets in July or enjoying the confines of a shed – will ever flourish. Would the public now accept this? Already sheep under relative intensification suffer a range of diseases, and in much of Europe, sheep farming is under a depressing cloud. Because of overproduction, the auction price of the animal carcass is so low that the whole industry is hardly viable.

Cattle and goats are very different from sheep. They have been easy to control, and over the last 5,000 years we have lived closely with these animals. Even today, in all continents apart from areas of North America and Australasia, there are examples of man literally living in the same building as his animals.

On my first visit to Switzerland in 1953 I was intrigued to see the cows, bells around the necks, being ushered into a communal barn – one end for the family, the other for the cattle. This is commonplace in villages in Africa, India and South America.

However, for several hundred years, herds of cattle have grazed in increasingly large fields in developed countries. In the last 20 years, the average size of a herd has become larger, and there has also been a move towards intensification, notably in the USA. In Europe, in most cattle rearing (veal production apart) the animals are free-range, at least for most of the year, and their artificial feeding has formed only part of their diet.

The intensification of American beef cattle rearing has resulted in substantial profits for the industry, and has made available an abundance of cheap beef on the international meat market. This is traded in most stock exchanges. Beef is usually transported deep-frozen, canned or chilled, according to length of journey.

It is not surprising that the USA is pursuing a policy aimed at reducing beef subsidies by governments for farmers in Europe, in order to promote its own highly competitive product. The end of these subsidies might make European beef too expensive to compete on the world market, with corresponding pressure to increase intensification and so lower the cost of production of its beef.

Milk

Cows have been bred to increase the amount of milk over the last century. A great deal of research has been done on milk and we now know the ways to increase the yield of milk maximally. First, the particular breed is important. Secondly, the feed: it is now thought that for greatest milk production cows must have access to more or less constantly available food, not just grass or silage, but additives rich in protein. Thirdly, the initiation of pregnancy must be carefully planned.

So the maximum amount of milk is produced by a cow under very carefully controlled conditions, and this explains the move towards more and more intensification. That is, the animals are housed more and more in sheds and barns.

Not surprisingly, so much milk is being produced in Europe that there is a vast store of unwanted milk – 'the lake', and butter – 'the mountain'. Despite this, there are moves to increase milk production even more by giving cows a regular 2-weekly injection of a hormone. This milk hormone is called Bovine Somatotropin (BST). The Food and Drugs Administration (FDA) in the USA has given permission in principle for farmers to inject their cows with this hormone, but the public is very cautious in accepting this practice. There is at present a moratorium on it in the European Community. The amount of milk could increase by about 15% and one of the inevitable results of this is that 15% fewer cows will be needed. Many small farmers will go out of business.

The hormone is very powerful. Unless the cows are reared artificially with a constant supply of feed, serious side effects are expected to occur. These include anaemia, infertility, infections, and defects of bone growth. How much pain the injection into the back gives is not certain – if only cows could talk!

The milk that is produced after the hormone is administered is different from the typical pint. The artificial milk contains changed fats, increased amounts of a secondary hormone called IGF-1 that affects the body's handling of fat and sugars, and increased amounts of the hormone itself. It is simply not known whether the artificial hormone can harm people, and it is not possible to perform realistic experiments to find out the long-term dangers.

Since we already have too much production of milk, surely we should not even consider using BST at all. Seemingly it can only harm cows and people. The beneficiaries would seem to be the pharmaceutical companies that make it. The drug is easy and now cheap to make through genetic engineering, and unless we are very careful, it will pave the way for a whole series of chemicals that no-one needs or wants, conferring benefit only to the manufacturers of the drugs. This hormone is the blue orange (Chapter 1) in reality; and consumers should continue to refuse to drink BST milk, and to demand that it be so labelled.

Global warming and cattle

There does not seem any doubt that the world temperature has increased to a small extent during the last century, and is expected to rise further, but by an uncertain extent.

Before butchers and farmers begin to think that I am the only (and misguided) critic of beef, another book gives a disturbing view of beef. The author is Jeremy Rifkin and the title is *Beyond Beef: The Rise and Fall of the Cattle Culture.*

We are given in this book a diet of health risks, corruption and ruthless exploitation of animals. But the real message is that as the world human population has grown, so has that of cattle, and many of the environmental problems are due not so much to human activity directly, but because we need so many cattle to support us: it is the cattle that do much of the damage.

There are now 1,280,000,000 cattle in the world, weighing more than the entire human population. Cattle are very inefficient feeders. For every 100 grams of plant protein fed to them, only 7 g ends up as meat protein. This explains why cattle need so much space, with the average being 1–2 acres for each animal. To give an idea of the scale of the problem, every beefburger eaten requires the destruction of 6 square yards of Amazonian jungle.

We all know that the major gas causing global warning is carbon dioxide, which is emitted by cars. But the gas methane traps 25 times as much heat as carbon dioxide, and 20% of methane in the air is belched out by cattle. So if we assume that global warming is a real problem, then the number of cattle currently being reared must make a major contribution to it.

The primitive atmosphere of the world, before higher animals evolved, was thought to contain large amounts of methane. Early plant life is thought to have generated oxygen that removed much of the methane. Are we reversing this process?

Pigs

Wild sows and boars were once found widely in woods and forests. Pigs are omnivorous, enjoying bulbs, tubers, roots, fruit and corn. It is their appetite for tubers and roots that causes them to reduce meadows to a muddy mess! Pigs also eat small mammals, if they can catch them, and they do have the justified reputation for consuming anything that they are offered.

In some ways pigs are among man's closest relatives. So similar is pig skin to human skin that pig tissues have been used in grafts in human surgery. The analogy with man is particularly important with regard to the nature of the fats in their tissues. Both man and pigs will store the surplus fat after it has been eaten. If either mammal eats a diet rich in sunflower oil (a good source of polyunsaturated fat), that same oil, if it is not used for energy, will be stored in the fat depots.

In the Middle Ages, pigs roamed wild in British forests, eating mainly vegetation, such as acorns. So then their body fat would contain mainly the 'desirable' vegetable oils. And when such pig meat was eaten by humans, those fats would become incorporated into our tissues.

So in this respect a pig is, like man, literally what it eats, and the problem is that were pigs to be fed on vegetable oils, the pig meat would tend to be oily, and so not popular with butchers and food processors. Thus bacon, pork and ham surrounded by firm, dry, white fat reflects the nature of the pigs diet. The composition of pig products is therefore partly determined for the convenience of the food processors.

Chickens

Most modern breeds of chickens are descendants of the red jungle fowl, which is still found in Java and Malaysia. These fowl would have perched high up on the branches of trees at night and foraged during daylight in clearings. They ate mainly vegetation, insects and worms, intermixed with grit or stones to aid digestion and make calcium available for shells.

During the last 100 years or so, particular named breeds were developed. Black Leghorn, Rhode Island Red, Buff Orpington, New Hampshire and the various types of bantams come to mind.

All these breeds of chickens have served both as a source of chicken meat and also for laying eggs. In the semi-wild, flocks were more or less contained with wire netting, and chicken houses were used for perching at night and for laying eggs. It would be the regular supplementation of their 'natural' diet, by putting feed into troughs on the ground, that would hopefully deter the birds from flying off. Despite the facilities provided, semi-wild chickens have a stubborn streak of laying eggs where they want to, rather than where the farmer wants them, and the chickens' wings are sufficiently well developed for them to fly some hundreds of yards.

A single cockerel would keep perhaps 10–20 hens content, with flocks managed to keep such a ratio of males to females. This meant that the superfluous males would tend to be eaten just before sexual maturity, and adequate fertilised eggs would be laid and allowed to be hatched by the broody hens to maintain the flock numbers.

All in all, this was a pretty haphazard and disorganised approach to producing food. It would be very inefficient by today's criteria. Some of the chicken available to eat would be the poor old hen at the end of her laying life, perhaps 4–6 years old. Yes, this meat was fairly tough, unless thoroughly cooked in the slow ovens of the day – as part of the range, the Aga or the Rayburn. But the flavour was delicious and well worth the trouble.

It is appropriate to introduce here the concept that so much of the flavour of meat is determined by the nature of vegetation consumed by the bird or mammal during its life. Indeed, this is not just a question of flavour. Chickens, like pigs, are omnivorous and their diet does generally determine the quality of the meat. Most people are familiar with the yellow colour of the corn-fed chickens, currently fashionable for reasons that are completely inexplicable! So, whilst difficult to quantify, it is common sense to consider that free-range chicken meat tastes more positively than that from intensely reared birds.

What about the eggs? The accusation 'grossly inefficient' at this system could well be justified. At the beginning of laying (incidentally, a cockerel is not needed to induce laying, just to ensure fertility), the eggs would often be small, rounded or mis-shapen, and would often break. When laying was established, the eggs would often be laid on the ground and sometimes trodden on, or covered with droppings. We were soon to learn that if such filthy eggs were washed, then bacteria could be im-planted in through the shell. Heavily soiled eggs were almost unsaleable.

Sometimes the eggs would get lost for days or weeks in under-growth before arriving stale at the retailer. Checks were made for freshness by observing the volume of air within the shell by transmitted light. The greater the volume of air, the older the egg.

There were further problems. The wretched chickens would stop laying altogether. Presumably this was a hangover from their origins, when at some time or other the chicken would have to take time off from egg-laying, and sit for a month on a clutch of eggs to keep them warm to hatch the chicks. In practice most breeds of chickens would lay, between December and March, about five eggs a week and then the number of eggs would dwindle during the summer. No wonder preserving eggs in buckets was a popular pastime for the farmer's wife!

So, based on today's efficient egg production, the system out-
lined above and in use over perhaps two or three centuries to the
1940s seems exceedingly archaic.

The main change in the current methods of rearing chickens
has been to separate the egg-laying capacity from those birds
bred for chicken meat. The latter are a different hybrid from the
egg layers, and the broiler, whilst capable of laying small eggs, is
usually slaughtered by 6–7 weeks of age, before it is sufficiently
mature to lay. Slaughtered even earlier are the 'poussins'. Such
broilers are reared in sheds and very much cheaper than the
traditional means of rearing chickens.

With both egg layers and broilers, a master or 'elite' breeding
flock is maintained, and from this grandparents and parents are
bred and can be transported across continents, either as chickens,
day-old chicks or as fertile eggs.

Geese

Fifty years ago, goose was a favourite luxury food at Christmas
and on special days. If you don't believe me, ask why there are
so many recipes for goose dishes in the old-fashioned cookery
books.

Why is goose rare today? Because I think it doesn't take at all
kindly to intensification and the birds have rather little meat for
their overall size. No doubt we intensively rear those animals and
birds that we can, but for geese we still have a great deal of
respect, as a result of their reluctance to do what we want them
to do (like pet cats!).

There are many species of wild geese such as Canada, Brent
and Barnacle. These have remained essentially biologically
unchanged over thousands of years. They reside in the cold
polar regions in the summer and migrate southwards in the
winter. Both wild and domesticated geese mate for life (unlike
many members of *Homo sapiens*) and have very set patterns of
behaviour.

Both the Greeks and Romans kept geese, partly as pets or

'guard birds', and sometimes to eat. Initially they tended to be red-breasted, but the Romans gradually bred the pure white type which laid more eggs than the wild geese. These pure white geese are the ancestors of most geese today. In ancient Britain, geese were often kept purely as pets, but in the Middle Ages some were reared for food. Because geese require water to flourish, the main centre for geese was the Fens, in Eastern England. In addition to food, geese provided feathers to fill the inside of pillows and mattresses.

Many of the Victorian painters of European farmyard scenes would include geese waddling in groups. With the decline of the mixed farm, geese are now the reserve of the amateur farmer, the smallholder and the enthusiast.

But these are birds that we respect. Not just their positive social behaviour, but their value in causing a rumpus when disturbed and their ability to convert rough grass to a fine sward of dense green turf.

Ostriches

The ostrich is the largest bird in the world, the adult weighing about 250 lb. and averaging 6 feet in height. Not surprisingly, these birds cannot fly, but they can't half run! Those of us fortunate enough to see these magnificent birds in parks may have seen them running at speeds up to 40 m.p.h.

Ostriches are long-lived, with the female laying giant, snow-white eggs for up to 40 years. The family unit typically consists of one cock and four hens (and I don't know what happens to the 'redundant' males). Ostriches are impressive wild birds that are biologically important, if not unique, and should command our complete respect. So why are we discussing ostriches here? Yes, you've guessed, there are moves afoot to farm them and eat them.

Ostrich meat enthusiasts point to its high-protein, low-fat content, and farmers say that rearing ostriches for food is more profitable than rearing cattle. Certainly in the wild, with the diet of ostriches being predominantly vegetation, the nutritional content of their meat should be good. Already ostriches are being farmed in North America, South Africa, New Zealand and Israel, and there is one farm in the UK, near Oxford.

If ostriches are to remain in wide open spaces, there will be few worries over their welfare, but the increasing intensification of deer farming provides a warning. Anyone endeavouring to farm ostriches in the UK has to be granted a licence from their local authority in order to comply with provisions of the Dangerous Wild Animals Act. It could just be that this government might maintain this hurdle in order to protect the beef farmers from competition!

But new breeds of birds, and cross-breeds, can easily be created by genetic engineering and it would be simplicity itself to create a hybrid bird, part turkey and part ostrich, sufficiently docile for shed rearing. The thought of this is quite horrible; surely we should leave birds such as ostriches as they are; and would it not be better and more environmentally sensitive for us to eat plant material directly, rather than feed it to farmed ostriches?

Other birds

Ducks and turkeys have all been reared for food since ancient times and much of the comment concerning chickens applies to ducks and turkeys. Today's turkeys are descended from a wild breed found in the southern USA and Mexico. There are in total at least 100 species of wild ducks.

The mixed farm

In Chapter 1, we discussed how soil fertility could be maintained by the use of manure from free-ranging birds and mammals. The point is this: even if short-term profits from free-range animals and birds are very much less than from those intensively reared, such forms of farming must be desirable as a means of ensuring soil fertility. Indeed, the combined arable/livestock system remained supreme in many parts of the developed world up to the 1940s.

Since that date the governments of those countries have striven to improve efficiency by the deliberate separation of these two types of activity. Indeed, the consumer has come to expect cheaper and cheaper meat – and has got it, not showing, until recently, concern over the way that meat was produced. With meat now an international commodity, in the same way as wheat or rice, once some countries embark on such policies of optimising meat yield, regardless of long-term consequences, other countries may also be forced to partake in this approach.

Britain has striven for greater and greater efficiency since the 1940s and more and more self-sufficiency. The loss of hedgerows and woods, and the damage to the soil are all well known. For decades the public was to remain largely unaware of the procedures used in intensive rearing. But now the cat is well and truly out of the bag. In the USA, the issues raised by intensive rearing have received less publicity than in the UK, mainly because so often farming *is* intensive rearing. There are no alternatives. It is now an appropriate moment to consider animal welfare.

Animal welfare

It is often difficult to define exactly when cruelty to animals does actually occur. We cannot tell what they feel, but we do know that in the main *our* morals have changed. We accept now a degree of cruelty in animals that I do not believe we would have tolerated 100 years ago. I can vividly remember appearing on a television programme, arguing against the use of Bovine Somatotropin (BST) to increase milk yield. To my amazement, a veterinary surgeon claimed that cows were happier being injected than not despite the large swelling that was produced on their hind quarters. I later discovered that this vet had been involved in the trials of BST, and therefore had a direct financial interest in the hormone.

No doubt our morals are influenced by financial gain, and indeed by defending a way of life. I cannot see how fox hunting is defensible, or hare coursing, or indeed the cause of any unnecessary torment of animals.

Animal welfare concerns have focused mainly on intensive rearing of chickens (broilers and egg layers), turkeys, pigs (particularly those chained in stalls), and calves grown for veal in tiny pens.

I am quite certain that if members of the public had been aware of what was happening to animals and birds as their rearing became more intensive, there would have been revulsion against the system 20–40 years ago. Today the expectation of cheap meat must reduce some of the natural opposition to the system.

A number of pressure groups are actively campaigning against such a system. In the forefront internationally is 'Compassion in World Farming'. In the UK, great credit goes to Chickens Lib. from Huddersfield, Yorkshire. Clare Druce, on behalf of this organisation, is rightly incensed by the battery system of egg production, which she thinks is illegal. Public pressure against this must surely mount, and it is gratifying to see that the phasing out of battery egg production is a policy commitment of the British Labour Party.

Vegetarians

More people have become vegetarian in the UK in the last 3 years than ever before. This must be partly a reflection on the publicity over intensive rearing and the feeling of horror of the unnecessary cruelty – all for profit.

Vegetarians are also increasingly secure in the knowledge that a completely acceptable diet need not contain any animal product at all. So let us look at the range of diet types.

A strict vegetarian eats no meat or fish or other products obtained at slaughter, such as lard or soup stocks. Vegetarians do eat eggs, milk and dairy products from live animals.

Vegans consume neither the mammals, birds or fish, nor any milk or dairy products, except of course human milk for babies.

Another major group of people has emerged, which to the author's knowledge is not identifiable by a simple slogan. These are people who do not eat fish, meat, eggs or milk produced under conditions of intensive rearing. So this would exclude battery eggs, barn eggs, most pig products, veal, and most ducks, and turkeys. Free-range chickens, pheasants, wild duck and deep-sea white fish would be acceptable.

Then there is a group of people who understand that fish do not possess a nervous system as we understand such a term and do not experience pain as mammals or birds might. They therefore do not eat meat but include fish in their diet, arguing in particular that fish is an excellent food nutritionally. Intensive farming of fish will be discussed in Chapter 4. Suffice it to say here that the world's natural stock of fish is sadly depleted, and it seems inevitable that further controls will be needed to preserve what is left.

3

Intensive rearing of mammals and birds – a true horror story

What is intensive rearing?

There is no simple or exact definition of intensive rearing. The systems used vary in different countries. It has also been inevitable that as the space required for the human population has increased, so that left for other purposes has shrunk. It is usual for most intensive rearing units in closed buildings to harbour animals (this familiar term usually refers to mammals) or birds closer together than with free-range, and also to be fed artificially. But there are further and probably more important differences between intensive rearing and free-range. These include issues of animal welfare, of profitability, of effects on the environment, and of the nutrients and safety of the final product.

The movement of animals and birds from open spaces to confined conditions cannot be traced to a single date. In any case, many animals, such as cows, inevitably spend some of their lives in enclosures, such as for milking. The provision of shelter for animals must go back to prehistoric times. Even today, sheep free to roam on upland grass must on occasion be grateful for shelter from winter snow.

Intensive rearing is not therefore principally concerned with the use of buildings in themselves. Rather, it is concerned with the maximum yield of meat, milk or eggs from the minimum resources. Here resources refer mainly to the cost of feed, of human labour and the investment in such facilities that enable optimal yield. It just happens that sheds usually provide this. Put another way, the intensive system is aimed at extracting all the possible usable material from the mammal, bird or fish.

The early part of this century saw a gradual development of intensive and also of partly semi-intensive systems in many countries. Initially, this included egg laying, production of veal and rearing of turkeys. The USA was prominent in these developments. However, in the UK, it was not until during and after the Second World War that intensive farming increased dramatically. There was pressure to grow more of its own food (in 1940 only about 40% of the national diet was grown in the UK) and there was a belief that the diet should contain meat and eggs. The scene was set for the route to cheap meat.

During the 1950s and 1960s, the American and British consumer's attitude to chickens changed from that of viewing them as an occasional luxury, to an essential basic and cheap food. The era of the meat and two veg., and three hot meals every day, was born. Everyone could afford meat for every meal!

It is only in recent years that this dogma has been questioned, partly as a result of the huge variety of ethnic restaurants in North America and Europe, allowing a much greater variety and freedom in eating.

The two central objectives in intensive farming were thus established: to increase the amount of meat available and to reduce the price. And of course, if looked at in isolation, these are laudable targets.

In both North America and the UK, these objectives can be summarised in just one word: 'efficiency'. Indeed, the striving for greater efficiency in food production has been the central preoccupation of our various governments for at least the last 40 years. Every aspect of intensive rearing has been analysed and researched. We know the optimum way of producing the final product, such as a wrapped chicken in the supermarket.

We know, for example, the optimum way of feeding the animals or birds. This is usually through providing solid food and water continuously or *ad libitum*. So to do this, large volumes of feed are placed in high containers (hoppers) with a gravity feed system including devices that ensure a constant trickle of feed to replace that consumed. We know that the physical form of the feed must suit the animal, and can be powdered meal, small grains or quite large lumps.

We know, too, the components in the feed that produce the most rapid weight gain, the largest eggs or the highest volume of milk. We know the importance of cereals such as wheat or maize as the basic feed ingredient, and that the addition of extra protein increases the speed of weight gain.

We know also the most efficient use of space in sheds. If animals or birds spend energy on movement, some of the beneficial effects of the tailor-made diet are lost. The optimum use of ideal heating and artificial lighting has also been carefully researched, as has the control of diseases.

We know also how to breed the species most suitable for the system, both in regard to tolerating the environment and also for putting on weight.

So the whole system of intensive rearing has been developed, improved and refined for optimum efficiency. In practice, the critical parameter with, say, broilers, is the speed of weight gain.

Consider a group of broiler sheds. The annual costs of their upkeep will be similar, regardless of how many chicken 'crops' are secured each year. This means that if chickens are ready for slaughter at 6 weeks rather than 9 weeks, then the productivity is increased by around 50%. Such increase in efficiency should result in a decrease in price for the consumer, increase in profit for the retailer, producer and middle men. Surely everyone is happy?

Well, not quite. The chicken is not over the moon, nor is the environment, nor is the consumer, when he finds out what is actually in the chicken, and nor are those responsible for public health. Let's look at some particular examples.

Cattle

The female of the species, the cow, supplies us with both milk and, at the end of her life, meat, with most of the edible animal going into burgers, pies, sausages, soups and so on. The typical cow lives to be 6–8 years old at slaughter, has four pregnancies, the last three effected by artifical insemination 3 months after the birth of her last calf. In the wild, the cow could be expected to live as long as 20 years.

The female calf is taken away from the mother after 24 hours; the young heifer is reared until pregnancy and birth of the first calf, and then for the last 4 years of the animal's life it is in effect a milk machine. It may be convenient in the summer for animals to graze in fields, but their feed is often supplemented and there is always the prospect of higher milk yield were the animals to be confined to sheds.

Certainly, some cows never see bright sunlight. The increase in milk yield (even without BST – see Chapter 2) may cause serious problems with the udder. Infection can occur. This is known as mastitis and will result in bacteria, pus and antibiotics contaminating milk. This milk should be excluded from the pool for human use and checks are made regularly to reduce the likelihood of this occurring.

The unnaturally swollen udder dangling between the rear legs sometimes causes an awkward stance and gait and this in turn causes disease of the legs and feet.

The male cattle are more likely to be reared entirely intensively, compared with the cows. Here the race is on to secure the fastest weight gain possible in the shortest time and hence command the highest price at slaughter.

Broilers

Most broiler chickens are housed densely packed on the floor of sheds. As the birds grow, their space becomes less. A shed might house 20,000–100,000 birds and a typical growing period is 6 weeks at which time the 'cropped' bird is 2½ to 3 lb. in weight. It is possible that this weight gain might in the future be achieved by 5 weeks. No doubt the UK Ministry of Agriculture is working on it.

Assuming 6 weeks of growth, plus a day or so for cleaning, a single broiler house may produce eight crops per year, or around half a million broilers. Well, not quite this number, because many will have died – say 5%, or 25,000. The broiler has been bred for maximum growth of meat, not for the strength of its bones or the vigour of its heart or blood vessels. It seems as if the bones and the legs just cannot support the weight of the bird, so some limp, collapse and die. Or perhaps the grazing bird's heart cannot cope and it dies of heart failure.

It is now routine for the house attendant to collect all the dead corpses on the floor on a daily basis. These corpses will usually be added to the heap of droppings accumulating outside each shed. The disposal of this chicken excrement poses real problems, and explains why incinerators may be needed just to dispose of chicken manure. The problem is that so many of these sheds occupy such a small space that there is insufficient land available locally to absorb the manure. On occasion chicken manure has been fed to cattle!

When freshly produced, excrement will have the pungent smell of ammonia, but gradually it becomes converted into nitrates that leach into our water. Nitrates in drinking water from East Yorkshire and Humberside in the UK exceed EC limits and are in part due to this.

The environmental impact of this type of operation can be considerable. Lorries are needed to deliver the feed and cart away the chickens. Some of the feed contains the bone meal fraction from rendered offal, including remains of earlier chickens, such as heads, intestines and feathers, all of which have to be conveyed to the feed compounders.

Bacteria such as salmonella would inevitably abound in the thousands of tons of droppings and rotting dead carcases which pile up. Vermin, notably rats, are attracted, and breed and multiply and spread them widely. Wild birds, too, feed off this debris, and there is the possibility – apparently within the law – of the excrement being fed to pigs or cattle.

Intensive production of eggs

There are basically three types of egg production, as seen in many European countries, under these cramped conditions. Most are laid by the battery system, where chickens are confined in tiny cages with sloping mesh floors. After an egg is laid it rolls into a collecting gulley. Feed and water are supplied continuously and the lights are switched on and off cyclically to give an imitation of day and night.

Other chickens are reared in large sheds or barns which contain perches and nesting boxes, but the conditions are still extremely intensive. This is the source of barn eggs.

The third type of intensive rearing involves thousands of chickens crowded on the floor of a shed with a tiny pen leading off. Occasionally, some chickens do venture outside, but many seem not to bother. There are regulations in the UK that enable eggs from these flocks to be called 'free-range'. This is a travesty of the real meaning of that word.

I prefer to call these eggs 'pseudo-free-range'. Because these chickens are the same hybrid as the battery, the eggs from them are just as likely to be contaminated with salmonella.

So some eggs labelled free-range in the supermarket are probably 'pseudo-free-range' and are no better than the battery. Indeed you, the consumer, have been the subject of a giant confidence trick, exploiting your emotions for concern over intensive rearing. It is hoped that in the future adequate labelling of eggs will become legally enforced. But until this happens, I suggest you buy an egg as truly free-range only if you can be certain

of the type of flock that laid it, preferably with your own eyes.

Incidentally, one reason why the supermarkets resist labelling eggs as to their exact farm of origin, is that this enables them to obstruct the tracking down of food poisoning and therefore reduces the likelihood of being blamed.

Vets

Recently, a leading veterinary surgeon claimed that free-range eggs were less desirable than the battery products. He then said that the organic beef herds suffered from more diseases than non-organic.

British vets have been making extraordinary claims over the past few years. First we heard 'there is no salmonella in any eggs', then 'salmonella is everywhere' and 'it comes from pigeons'.

May I suggest explanations as to why some vets (probably a very few) feel obliged to mislead the public? First, it is guilt: they have condoned cruel and dangerous farming practices. Secondly, intensive rearing must be associated with a greater incidence of disease than free-range husbandry, so the farming vets have a direct financial interest in these practices, as do the companies that provide the required drugs.

Salmonella

To return to salmonella, it has been known for years that modern intensive methods of broiler and egg production are addled with salmonella. The UK Junior Minister of Health, Edwina Currie, *was* right, even though unpopular, when she said in 1988 that most of our egg *production* (my italics) was contaminated with salmonella. In my laboratory in the UK, in the summers of 1991 and 1992, we have identified more food poisoning bacteria from ill patients than ever before. The problem is not solved. Nor is it just a British disease.

In recent surveys from Georgia, California and Carolina, USA, the extent of the salmonella contamination of intensive chicken and egg production is disturbing. Sixty per cent (yes, 60%) of samples of recycled chicken protein going into the feed contained salmonella. Salmonella was found in one in 10 of hatching chicks, 5% of broiler droppings, one in three samples from trucks used for carting live birds, and one in five of all broiler chickens. In 32 out of 42 (76.2%) of flocks, salmonella was found in the ovaries from laying chickens. It is this contamination that causes salmonella to get into the egg yolk, and is the reason for the advice that eggs should still be cooked until the yolk is hard.

Artificial feeds

Most artificial feeds are cereal based. These cereals include powdered maize, wheat or barley and may be imported. The cereal will need to be compounded into a formulation with added vitamins, minerals, protein supplements and drugs to produce the right nutrients and flow properties for the hopper and feed systems. Feed compounding is a skilled process and there are relatively few companies operating.

The resultant feed from one central source can therefore be distributed very widely. This type of production is potentially very dangerous, were anything to go amiss, such as the introduction of an infectious agent.

It can, however, be argued that because of careful monitoring and controls, and that test batches of the feed can be shown to be safe, therefore by inference the whole output can be guaranteed. Whilst fine in theory, the problems in practice relate to possible unknown (and therefore not capable of detection) dangers.

Rendering plants and Bovine Spongiform Encephalopathy

With each substance added to the feed, so the risks multiply. In most countries the high protein concentrates are derived from the rendering plants. What are rendering plants? Some parts of the intensively reared animal or bird are inedible. These include bones, feathers, heads and feet, and intestines. These products are processed in rather smelly and secret factories into two usable products. Fat or tallow is used for making soap, lipstick and other cosmetics. The other component is a protein-rich material that is further processed. In the past this has been referred to as bone meal and used as a garden fertiliser.

Exactly when the practice of incorporating this type of material, now referred to as concentrates, into animal feed began is difficult to be certain about, because of the surrounding secrecy. It was probably in the late 1960s or early 1970s in most developed countries. The appeal is obvious. The unwanted waste material can be converted to a useful and profitable industry. When such feed is returned to the same species as that from which it was obtained, it is inevitably a type of cannibalism, in the same way as the feeding to pigs of household bacon scraps.

The rendering plant operators, the feed compounders and the

farmers would not have been aware of any infectious risk in this system. Surely the temperatures in the rendering plants would have been adequate to kill any germs?

With hindsight, we now know this is not the case. Unfortunately the infection that causes the cattle disease, Bovine Spongiform Encephalopathy (abbreviated to BSE and dubbed Mad Cow Disease by the media) is resistant to the treatments in most rendering plants. This disease has already destroyed at least 100,000 UK cows and has spread to Switzerland, France and Denmark. It was thought initially to be spread by animals' organs being fed back via the rendering plants (for further information, see Chapter 9). Fortunately, it has not appeared in the USA, but there is no reason at all why it should not in the future.

This means that the whole system of interdependence between the rendering plants, animal feeds and intensive rearing is unsafe. Oh dear! Already the UK government has prohibited the use of these concentrates being fed to ruminants, but strangely not to pigs or poultry.

There is another skeleton in the cupboard. There are 7 million cats in the UK. About a million die each year. Many dead cats are disposed of through the rendering plants, after some of the skins have been sold for furs. Not very appealing? But it's worse than that, because the practice is dangerous. If you are still reading, let me try to explain.

The first cases of BSE of the present epidemic occurred in 1986. The first case of the similar disease in a cat called Max in Bristol, UK was reported in 1990, presumably after eating contaminated beef products. Many cats are thought to have succumbed from the disease, which we will call FSE or Feline Spongiform Encephalopathy. So, if these dead, infected cats (many may be carrying the infectious agent but die of other causes) are treated in the rendering plants, then their infectious agents will be added to the feed of animals such as pigs and also to poultry and then to us.

Of course we do not know and cannot tell for certain whether human beings can be infected with FSE, although we do know that the human dementia, Creutzfeldt–Jakob disease, can be passed experimentally to cats. There is no reason why the infection cannot pass from cats to man.

As the law stands, I think, in all countries, it is perfectly possible to dispose of dead cats in the rendering plants. It is interesting that it is now not permitted to treat certain organs of cattle in the plants in the UK. An immediate total ban on this means of disposing of deceased cats (and other pet mammals) must be introduced urgently. Instead, they should be incinerated. We will need to build more incinerators anyway, for the livestock remains that cannot now go to the rendering plants. You probably did not want to read this, but most of you, particularly in the USA, have been protected from this information. The debate will have to start. This is very bad news for the beef industry in the USA.

Drugs added to the feed

Drugs are added to the feed (and water) of intensively reared mammals and birds for two purposes: to treat and prevent diseases, and to increase their growth rates. It is the latter that gives the ethical worries.

Antibiotics are by far the commonest feed additive for both these purposes and doubts over much of their use continue. Antibiotics are usually produced by microorganisms in tiny amounts and the pharmaceutical industry has become expert at concentrating and purifying these so that they can be used for treating infections in people (and animals). Some antibiotics in use now are chemically altered natural products and some are completely novel chemicals.

Unquestionably, antibiotics have revolutionised many aspects of human medicine. For example, meningitis, typhoid, tuberculosis, osteomyelitis and pneumonia all can be treated as can less dramatic but annoying infections such as thrush, urinary infections and septic skin spots. However, it is probably in their use associated with modern surgery that their impact is most striking.

Soon after antibiotics were first used in human medicine in

the 1940s, some bacteria were identified that had previously been susceptible or sensitive to an antibiotic, but had become resistant to the effects of the drug. Resistance in bacteria to antibiotics does seem inevitable to varying degrees, and it is the amount of drug used that seems to decide how many bacteria become resistant – the greater the use, the more the resistance. Indeed, many experts now consider that some of the use of antibiotics in human medicine is unnecessary, such as for most sore throats, and doctors are being encouraged to use antibiotics only when really needed.

Many infections are caused by bacteria that inhabit both animals and man, e.g. salmonella, and it is now fairly obvious that if an antibiotic is used in animals, some of the animal's bacteria may become resistant and those bacteria can then spread to man. This spread can occur by direct contact with the animal, by touching meat, by air, dust, flies, and other insects.

So, many authorities believe that the use of antibiotics in intensive rearing can cause difficulties, increased costs and even danger for the treatment of infections in patients. However, proof of this is difficult to come by, and the makers of the antibiotics in animals do have a good defence: they claim that most resistance in human bacteria is due to the high and sometimes unnecessary use of antibiotics in man.

However, whilst the exact amount that antibiotic use in animals contributes to resistant bacteria in human infections cannot be quantified, surely all agree that the reduced use of antibiotics in food animals and birds would be a laudable aim?

This will be difficult to achieve, for two reasons. The first is that intensive rearing, by definition, must inevitably pack the birds or animals close together, which results in the easy transfer of bacteria and viruses between individuals, either through direct contact or in the very short journey through the air. One animal might contaminate the drinking water or feed and so transfer the infection to others. We know that in human populations, the closer we live together, the more likely we are to suffer from epidemics of infections spread in the air, mainly infecting the lung.

What this means is that intensive rearing itself must inevitably increase the spread of infections amongst animals and birds, and antibiotics will be needed to control the bacterial causes.

But the use of antibiotics to stimulate weight gain can and should be stopped, whether or not intensive rearing continues. It was by chance that livestock, having been inadvertently fed the antibiotic tetracycline in the early 1960s, put on more weight than expected. This occurred in the USA and much confirmatory work done there establishes that long-term use of certain amounts of antibiotics in the feed of most intensively reared species causes an increase of weight over several weeks. The gain is not dramatic, perhaps 3–5%, but this is sufficient to be decisive commercially.

This means that if one farmer uses growth-promoting antibiotics, then others may be forced into the practice. It has continued because the pharmaceutical industry has been able to use its powerful influence to defend it; surely any common sense or natural law would indicate it to be undesirable. In many ways this encapsulates everything amiss with the system of intensive rearing of animals and birds. Profits before Ethics.

Other drugs are also added to animal feeds, for treatment and prevention of disease and to increase growth. Most drugs are carefully tested and careful calculations are made to identify the time needed for the drug to be removed from the animal's body, usually through its excreta.

This means that animals or birds treated with drugs should not be slaughtered and subsequently consumed until adequate time has elapsed for that elimination to have been completed. Farmers and vets should know this length of time (the withholding period) and should comply with it precisely.

But the need for meat is not always predictable, and there will undoubtedly be pressures resulting in the animals being slaughtered before the drug has been eliminated. For this reason most developed countries undertake a regular test scheme for drug residues in meat. This is in effect saying to the intensive farmer, 'We do not trust you!' Amazingly, the UK government has admitted that in 1990, 5% (yes, 1 in 20) of samples of pork tested were tainted by the antibiotic sulphonamide.

Clenbutarol

Clenbutarol is a drug that has been used illegally in several European countries and its presence has been detected in many samples of beef. The drug is known colloquially as Angels' Dust, and a secret organisation for its distribution and use has centred on Holland, France, the UK, and Ireland.

The drug has been imported in bulk from South America and distributed by using falsely labelled medicinal containers. Farmers have sprinkled it on cattle feed and the effect is to convert some of the animal's fat tissues to lean meat. The value of the carcass at slaughter increases – indeed, it may rescue an effectively non-saleable commodity.

There is evidence of serious illness induced by the drug in cattle and the potential for damage to human health is considerable. The drug is known as a beta-agonist and has powerful effects on the lungs, heart and circulation. One effect is to counter the effect of drugs used for treating high blood pressure or irregularities of the heart. Such drugs are called beta-blockers. It is certainly possible that the eating of meat containing even small amounts of Clenbutarol could result in serious symptoms, or even sudden death in these patients.

It is gratifying that the Irish Government is taking firm action to try and stamp out this illegal practice.

The Clenbutarol issue has shown that intensive rearing of animals and birds is easily exploitable by illegal operations for short-term financial gain.

It is not intended to imply that this practice is at present widespread. Indeed, the majority of farmers must be behaving within the law. The problem for the consumer is that he or she is unable to identify the particular origin of most meat available, even that which has not been subjected to processing. Incidentally, it is this same drug that was used by some athletes in the 1992 Olympics to increase the bulk of human muscle.

Blue ear pig disease

Blue ear disease in pigs is a type of influenza and its occurrence must be encouraged by intensive rearing. Sometimes the lungs are badly affected so that oxygen cannot get into the animal's blood. This makes the blood look bluish, particularly in the ears.

The recent epidemic in Europe began in November 1990 in Germany and has spread westwards, with hundreds of thousands of pigs recently infected. Human influenza spreads similarly. Every 10 years or so, a new type of the virus spreads among the human population, beginning in the Far East and then travelling west. Hong Kong 'flu was a typical example. It is thought that these waves of the human disease originate in pigs in the Far East and Central Asia. Much research has been performed explaining exactly how the influenza virus keeps changing its properties; this is known as mutating. The vaccines against human influenza give some protection against the disease in people, but as with all vaccines, they have no effect on other diseases.

In 1991, the researchers in Holland discovered the virus that causes blue ear disease in pigs. They also found that if human influenza vaccine is injected into pigs it can protect them against blue ear disease. This means that pig influenza is very similar to human influenza. But is this important? The answer to this question is yes, because the disease in pigs has a particular tendency to infect pregnant sows and destroy their piglets. If this occurred in women, there could be a similar problem. Indeed there are examples of other infectious agents causing disasters for both pregnant women and animals. Listeriosis can destroy human babies and also lambs and calves.

At the moment, all the features of the blue ear virus are not known, nor is it established whether people have already been infected with it. One major problem is that human influenza strikes mainly in the winter, and there is no way of telling how many cases of the disease in Holland and Germany were due to a human or the pig source. So no relevant research has been performed studying the transfer of the infection from pigs to people. It is certainly possible that some people might have been infected, perhaps without showing obvious illness. They could then pass the infection on to others.

Were pregnant women to be vulnerable, the effects would be unpredictable. At best the blue ear virus would not damage the baby; at worst it could produce birth defects, as can german measles. It must be common sense for pregnant women to avoid contact with live pigs from infected herds. This advice is described as 'scaremongering' by the pig industry. But may I remind the pig farmers that it is their method of intensive rearing that was the cause of the virus spreading so rapidly amongst pigs.

The battery egg

It is probably the revelation as to how most of our eggs were laid that disturbed people most, regarding animal welfare over the last few years in the UK. The battery system does represent the ultimate approach to efficient food production all over the world, and has been developed seemingly regardless of any real consideration for the welfare of the birds. One wonders why organisations dedicated to animal and bird welfare have in the past not been more active in criticising the system.

Maybe our attitudes are influenced by the final product. The egg shell is usually smooth and regular, with no sign of droppings, feathers, cracks or damage: a perfect food package. Also, the egg is usually fresh in the real sense of the word. Some egg producers can actually ensure delivery to retailers within 24 hours of the egg being laid. Freshness is appreciated by the consumer as a firm and uniform-textured albumen (or white), the yolk sits up proudly in the frying pan and the golden-orange hue reminds us of the sun. Most of the egg delivery lorries will describe the battery sheds as a farm. On them will be painted delightful 19th century rural pictures of chickens scratching around in meadows, the surrounding oak trees casting shadows from the golden sunset. The packing station boxes will usually reinforce this idyllic illusion. The consumer may not appreciate, nor want to know why, on trips deep into the country on holidays, he or she sees no hens in the field, and may not be curious as to what is going on in those giant windowless huts with strange projections on the roof.

Then nutritionists say, correctly, that egg albumin is an excellent source of protein; and whilst the consumer may have heard that high amounts of cholesterol are present in the yolk, there is so much confusion over cholesterol, no-one believes the experts anyway; and surely, if it comes from a farm, then it must be healthy.

Then, finally, there is the price. Eggs are cheap, several-fold less than the equivalent price in 1950, and of course they are always available, every day of the year.

Is not the modern egg the answer to the consumers' prayers? Surely it is better than the expensive, dirty, damaged offering, often in short supply, from the bad old days?

What actually goes on in the battery egg system?

Over the last 40 years, chickens that lay battery eggs have been carefully bred to lay continuously (i.e. nearly daily) for about a year; the eggs will be a large and uniform size, and the chicken will be as efficient as possible in converting its feed into the egg.

Amongst the layers, the problems of diseases are controlled essentially by drugs and vaccines, and disturbed social behaviour such as pecking at neighbours is managed by clipping off the beaks. The female breeding or 'elite' hens are provided with adequate cockerels to lay fertile eggs. On hatching, these represent 'grandparent' stock and are regularly sold to breed 'parents'. It is from these parents that fertilised eggs are transferred to the hatcheries and after incubation, the day-old chicks emerge.

Hatching day-old chicks in isolation from their mother (or even father) provides the most efficient form of breeding; but there is a problem. Most newborn birds or mammals become colonised by beneficial bacteria from their mother shortly after birth. These bacteria persist for the whole of the animal's life and their main value is to keep possibly dangerous germs at bay, simply by their sheer numbers. This does, of course, happen with human babies: the baby in the mother's womb will contain no bacteria, but shortly after birth, the baby will harbour the mother's bacteria from her skin, birth canal, mouth and so on.

But not the newly hatched chick – there is no mother in attendance. This absence of the normal bacteria from the mother hen is thought to be one of the reasons why salmonella is such a problem in poultry. Research has shown that if these normal or helpful bacteria are given to the day-old chick, it will resist salmonella.

Because salmonella may be on the membranes covering the chick as it hatches, the day-old chick may well become invaded by salmonella, and even die; but more often it recovers subsequently, with the salmonella still living in its internal organs, later to infect eggs or contaminate broiler meat at the time of slaughter.

When day-old chicks hatch it is easy to distinguish the males from the females. The males are usually killed by gassing, and their corpses carried by conveyor belt to a type of reprocessing unit, so their remains can be used as feed. The owners of the larger broiler/egg laying units claim nothing is wasted! How right they are.

The females (a few males do survive by accident, and are subsequently weeded out), having been delivered as day-old chicks although they may be somewhat older than this in practice, are then reared intensively, usually on the floor of sheds. They are transferred to the broiler cages at the point of laying, say at 18 weeks of age.

A typical cage for five hens measures 18 in. × 20 in. (45 cm × 50 cm). There is a supply of piped water and meal, so the chickens are fed *ad libitum*. The cages are stacked in rows of about 3–12 and in long lines, with a large broiler house harbouring many thousands of chickens. The floor of each cage consists of a wire network and slopes slightly so that eggs after laying roll gently into a trough away from potential danger from the feet and the droppings. This trough can be static with eggs collected manually, or it can be part of a moving conveyer belt.

Droppings fall through the wire mesh and are regularly removed. The internal environment of the shed is usually controlled for temperature and humidity. Lights are automatically switched on and off cyclically to mimic night and day, but of course the chicken has no notion of seasons: if it were to believe that summer was approaching, it might stop laying!

The major problem in the battery is of chickens pecking one another – not surprising, as this is just about all the entertainment that is on offer. So the typical battery chicken has scruffy looking plumage, it may have some completely bare skin, it may bleed and its behaviour can become so aggressive that bones are broken and it dies. On a daily basis each large house will have each cage inspected and any corpse removed.

After about a year, the chicken tends to lay fewer and sometimes misshapen eggs. The crop is slaughtered and the miserable carcasses used for processed chicken products, such as soups, pies, pet food and 'burgers'.

Let us return to the eggs. These will either be packed on site or transported to a packing station. Here eggs are inspected both visually and using light to illuminate the egg's centre. Defective eggs are still used, for example in the cake industry, or for making dried egg powder. Nothing is wasted. The remaining eggs are graded according to size, and packed into containers with dates indicating recommended maximum times of storage.

One final point about the colour of the egg yolk. Much of the golden-orange colour is produced by putting suitable dyes with the chicken feed, in the same way as farmed salmon are pinked by dye in the feed (see Chapter 4).

The benefit of the battery system to the egg

The traditional bad egg is on the way out and the battery system is partly responsible. By the bad egg, I refer to eggs which are obviously discoloured, brownish, grey, green, black or even pink. The yolk and white merge into one another and the sulphurous stench is ghastly.

These effects are due to bacteria gaining entry to the inside of the egg through contamination of the shell surface with droppings or soil, and damage to the integrity of the shell through water or cracks, usually with rupture or penetration of the membranes between the inside of the shell and the albumen. This can be caused by accidents on the ground or the hen treading on the egg. The battery system permits the egg to roll out of harm's way soon after the egg is laid.

In conclusion, the genuine and properly husbanded free-range system must surely be the preferred type and surely eggs are so cheap that the moderate additional premium is justified, notwithstanding the rare bad egg!

4

Salmon and polyunsaturated fats

Salmon in the wild

Wild salmon are handsome, biologically important and in many aspects unique among fish, inhabiting alternately the sea and fresh water. Until recently, to eat salmon was an expensive luxury. Smoked salmon, caviar and champagne were for the very rich or for very special occasions!

Because salmon in the wild move between fresh (or river) water and the sea, their natural habitat in the sea is relatively near land. The Atlantic salmon is found down both sides of the Atlantic. In the European margin, they are found from sea areas adjacent to Northern Russia to the French and Spanish border, and also around Iceland and Greenland. Along the western waters of the Atlantic, they flourish from Hudson Bay in Canada to Cape Cod off Massachusetts in the USA. The Pacific species are distinctive and in general found in sea water not too far away from land in northern latitudes.

The life cycle and feeding habits are complicated but they do account for the distinctive nutritional content of the mature wild fish. The young salmon hatch from fertilised eggs laid on the stony beds of rivers: in the UK these are mainly in Wales, Scotland and Ireland. After hatching the tiny fish keeps with it a bag of nutrients from the egg. This is called the yolk sac and is attached to the under or ventral part of the fish. The yolk sac provides enough nutrients for the baby salmon to establish other sources of food in the river.

When the fish are 7 cm long, they are referred to as parr; and at 15 cm as smolts – they will now be 2 years old, silver-coloured and ready for their excursion down river to the sea. The early growth of the salmon can be slower than this, and in cold water some may not be large enough for their sea migration until 5 years old.

In the sea, the salmon feed on small organisms, which themselves have fed on even smaller organisms and so on. Much of marine life is ultimately derived from the tiny phytoplankton, and it is from their ability to synthesise substances such as polyunsaturated fats that these and the pink colours become concentrated in the now rapidly growing salmon. Research has shown that one factor favouring these desirable or even essential fats (i.e. we cannot live without them for long) is a relatively cold environment.

After 2–3 years in the sea the salmon are now mature and the males and females return – quite incredibly – to the upper reaches of the same rivers from which they hatched. Some theories suggest that salmon recognise 'their' river through smell, but the mechanism is difficult to understand. There is evidence that if the mouth of the river is polluted by, for example, farmed salmon, or is muddy, then the salmon may become confused and fail to find their breeding grounds. Whatever the mechanism, it is a truly amazing and biologically important phenomenon. As the salmon move upstream, they tend to live off their fat stores, rather than attempt any serious eating.

Whilst it was fairly easy for the small parr or smolts to swim downstream with the current, the larger adult salmon has a very much tougher assignment, particularly where the river cascades down over boulders. So the salmon jumps, using its well-developed tail muscles to generate violent movements of its fins which propel its body up through the water and even out of it altogether through the air. Unfortunately for the salmon, this makes it very conspicuous to the fisherman!

The salmon breed in fairly shallow but free-flowing water over gravelly ground. After an initial courting ritual between the larger male and the female, it is the latter that makes a hole in the pebbles by a sweeping movement of her tail fin. She then lays the eggs in the hole, the male fertilises the eggs and it is the female once again responsible for filling in the hole with gravel.

After this activity the salmon are exhausted and begin to swim downstream, where they are a fairly easy prey for man. However, some do reach the sea once more and then return and breed in subsequent years. It is obviously preferable, therefore, for the survival of the species, that adult salmon are caught on their journey downstream back to the sea, rather than upstream to breed, and the silver smolts spared completely.

The nutrients of wild salmon and other fish

The pink flesh of wild salmon is composed mainly of protein, as are the muscles of all organisms. It would seem that because of their singular determination to migrate upstream to their breeding ground, their muscles have to be so well developed. There is overall rather little total fat, and certainly almost no saturated fat in wild salmon. Typically the total fat is 3% of the fish by weight. What fat there is, tends to be a particular type of polyunsaturated, known as n-3, ultimately derived from the plankton in the sea.

The realisation of the importance of these fatty acids came from the study of Eskimos in Greenland, 15 years or more ago. Eskimos very rarely suffer from heart attacks because their arteries do not tend to accumulate cholesterol deposits and clots. The freedom from these was in turn attributed to large amounts of n-3 fats in their cold, fishy diet. There is also evidence that n-3 fatty acids are desirable for human brain development. Fish oils are thought to be beneficial in preventing or treating a host of other diseases, including arthritis, eczema, psoriasis, disorders of immunity and even cancer.

In wild salmon, it is their diet of natural sea organisms, coupled with their active movements and the low temperatures of their environment, that produce such a remarkable food.

Farming of salmon

Artificial farming of salmon has been undertaken in Scotland,
Ireland and Norway for more than 10 years, and has been beset
with criticism, environmental concerns and now doubts over
nutrients and the safety of the product. In addition to salmon,
the farming of trout is well established, and other species are now
going through the experimental stages, such as halibut. Farmed
salmon is, however, the major industry and has been most studied.

To farm salmon artificially, it is necessary to re-create the
conditions suitable for its development. This means that the
majority of growth must occur in sea water, which explains why
the holding nets are located in sea lochs or are in inlets artificially
constructed from the sea. For success the temperature of the
water must be on the cool side. But before entering the sea nets,
the salmon must be reared in fresh water, and the whole process
is performed in a series of large tanks or growing chambers,
beginning with artificial fertilisation of eggs from breeding
salmon.

As the fish grow from parr to smolts, so they are transferred
between tanks, there being a flourishing trade in these little fish.
Two problems remain: one is the provision of the feed; the other
is the prevention and treatment of diseases associated with a very
high density of growing fish. Entire batches of young fish can be
lost because of disease.

When the growing salmon are transferred to their sea environment, they are checked for disease, but still many problems
continue. What are the growing salmon fed with? It is impossible
to recreate the natural sea diet, but attempts to get as near as
possible to this have been made. A common source of feed is
small fish from the Norwegian Sea, in powdered form. But other
products can be used and the farmed salmon have a voracious
appetite and will take to a huge variety of feed given to them.
One salmon farmer told me that his last batch of feed had some
chicken feathers in it! I'll leave the implications of this to your
imagination. Even recommended feeds can contain 10% beef
liver or 25% cereals.

Pinking farmed salmon

Readers will be aware that many sea creatures – shrimps, crabs and lobsters, for example – are pink. This pinkness is due to natural dyes in the tiny creatures on which they feed. Wild salmon is also pink. But farmed salmon are pink only because an orange-pink dye is incorporated into their feed. Without this, the colour would be anything from off-white to dirty grey. Colour provides important expectation when eating food, and there is evidence that colour plays a major role in influencing whether animals or birds eat a particular type of food. This is partly hereditary or instinctive, and partly as a result of learning. In practice, then, the consumer would not accept white or grey salmon, hence the need for incorporation of a dye in the feed.

Until recently, the dye mainly used was a synthetic substance called canthaxanthin. It is similar in chemical structure to Vitamin A. The group of these chemicals is referred to as retinoids, and it is necessary to discuss them a little further. Vitamin A itself has been known for decades to be essential for several body functions such as seeing at night. In order to ensure adequate intake of this vitamin, some foods such as margarine must contain the added vitamin: they are then said to be fortified. Incidentally, the orange colour of carrots is due to a substance called carotene, some of which is converted in the body to Vitamin A.

It had been thought that the effect of Vitamin A was entirely beneficial, but there is now evidence that eating too much of this vitamin or some related chemicals can be dangerous. The main effects have been shown to be in pregnant animals; in humans this can result in birth of offspring with deformities of the face such as harelip and cleft palate. Indeed, the British government in 1990 issued a warning to pregnant women not to eat liver. This was quite unprecedented because liver had always been thought to be an excellent source of vitamins, such as folic acid, Vitamin B_{12} and Vitamin A. The reason for this warning was that testing had shown dangerous concentrations of Vitamin A in liver.

Incidentally, a drug available in the USA and the UK for treat-

ing severe acne is also chemically similar to Vitamin A and as
such is referred to as a retinoid (meaning relevant to the retina,
the back of the eye). Female patients treated with this drug are
advised that they must not become pregnant.

By 1990, the problem with canthanxanthin was becoming obvious. It was a retinoid too; so what was the safe amount to eat?
No-one knew! In particular, virtually no experiments had been
done in experimental animals, and in any case would those results
be relevant to dangers in people? Again, no-one knew!

In practice, it is impossible to test safety adequately for such
a chemical over a short period, such as 2 years. Another problem
became apparent in 1990. Canthaxanthin had not been cleared
in the UK as a colourant to be added to food directly, but it was
being used – apparently legally – as an additive to the feed of
laying chickens to darken the colour of egg yolks, and also as a
feed additive for salmon.

So a difficult dilemma had developed for the British Ministry
of Agriculture, Fisheries and Food (MAFF), made more acute
when it was publicised that two of its scientific advisory com-
mittees recommended – quite logically – cessation of use of canthaxanthin in feeds. But what did MAFF do? Did they ban
canthaxanthin? No, they did not. In a move that can only be
described as brilliant (tactically), because Norway was also using
the dye, MAFF referred the general matter to the European
Commission.

Meanwhile, the salmon farmers were encouraged (but not
forced) to replace canthaxanthin with astaxanthin, a similar but
natural dye.

There are many people who wish to identify the source of the
salmon, as to whether it is wild or farmed. Much of the salmon
sold in the UK is labelled simply 'Scottish salmon', which is
presumably farmed. The colour is a good guide. The farmed
salmon has often been fed too much of the colourant so that it
is too dark an orange-pink colour. Wild salmon, such as Pacific,
tends to be a paler mid-pink, or just orange-tinted.

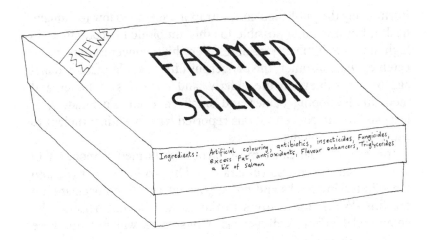

Drugs in salmon farming, the environment

The next problem with farmed salmon concerns the use of drugs in its rearing. As with intensively reared poultry or mammals, their high density encourages the rapid spread of diseases. These include bacterial and fungal infections and insect infestations. Drugs are available for treatment, and these are usually well researched and authorised by a veterinary surgeon.

But the problems of drug use are greater than with, say, poultry. How do you prevent the drug spreading in the water and damaging marine life on the sea floor? The fish needing treatment can be transferred to a chamber to be treated with the drug being put in the water, and then returned to the confines of the net within the sea. But the drug might then leach out of the fish and damage the environment. This means of drug delivery is expensive, particularly when many treatments are needed.

Alternatively, the drug is sprayed on the top of the water over the mass of wriggling fish in the hope that an adequate amount will get into the fish. It seems that however the fish are treated, in practice some spread of the drug is inevitable.

One antibiotic often used for treating bacterial infections is tetracycline, and there had been well-documented reports of salmon at the point of sale being contaminated with this antibiotic.

Fortunately the published concentrations seem too low to damage health, but it is quite possible for this antibiotic to accumulate to high amounts in nearby wildlife. Were high concentrations to be eaten by the consumer, the dangerous effects could include damage to the kidneys in the elderly, and disorders of bones and teeth in developing babies and small children. Fortunately, at least so far, the concentrations reported have been insufficient to cause these problems.

It is sea lice that cause most havoc with farmed salmon. (Wild salmon seem much less vulnerable.) The sea louse burrows into the skin leaving tracks and holes, particularly on the back and on the fins. Whilst the distress to the salmon is considerable, the commercial effect is disastrous. Consumers will not purchase obviously disfigured fish.

Over the years concerns over the persistence of insecticides such as DDT have caused a shift towards those drugs that tend not to persist for long in soil or water. The chemicals now in widest use are called organophosphorous compounds. For example, they are incorporated in solvents or shampoos for the treatment of head lice in people. They kill insects after contact through paralysis and they can, if absorbed in people, be dangerous to the point where they can produce immediate death or, if a person is exposed for weeks, a chronic disease of confusion, shaking, weakness or tiredness. For serious side effects to occur the drug would either require to be eaten or applied to a very large part of the body surface.

The only effective compound used for delousing farmed salmon is one of these organophosphorous drugs, known as dichlorvos. The drug is added to the water at intervals, and the lice are killed by it. It certainly does work. However, the salmon can become re-infected and several applications of dichlorvos may be needed.

Occasionally a salmon will absorb excessive amounts of the drug – so much so that it is killed. This is unusual, but whilst the interval between the last use of the drug and the retailing of the fish should be sufficient for the drug to have disappeared from the fish, this is not always the case. Salmon in one supermarket have been tested and found to contain dichlorvos. However, the more serious side effects of dichlorvos occur in people handling the drug, who could accidentally absorb amounts sufficient to be very dangerous. There are reports of actual deaths in people associated with using organophosphorous compounds.

The effect on the other marine life has also been shown to be damaging. Whilst dichlorvos persists for shorter periods than older insecticides, its rate of decomposition varies with the temperature and acidity of water, and cold sea water does permit the drug to survive for days or even weeks in sufficiently high concentration to kill many marine organisms. Sea 'deserts' have been described near salmon farms. Incidentally, the pink dye from the feed can also escape, and then animals that survive dichlorvos can turn bright orange-pink!

The list of environmental concerns with farmed salmon goes on. The avidly eating conglomeration of fish produces an enormous amount of excrement, which drops to the sea or loch floor. There, further chemicals are produced by decomposition with damage to marine life. The impact of this can be lessened by movement of the farm at regular intervals, but this adds to the cost and it is arguable that this might just spread the poisons more widely.

Another environmental concern is the possible effect on wild salmon. Here there is no formal proof, but a very high degree of suspicion. Articles on fishing and conservation during the last few years have pointed to a reduction in the numbers of wild salmon inhabiting rivers. The opposite trend might have been expected since salmon farming, through producing an abundance of cheap salmon, should have reduced the commercial pressures on the costly and unpredictable wild salmon trade.

There are many salmon fishers and environmental commentators who believe that farmed salmon have directly been responsible for the loss of wild salmon stocks. The argument goes as follows. The method of rearing farmed salmon has created a type of fish adapted to this artificial environment and not suitable for wild habitation. It has been selected to put on weight rapidly; and it will not need natural conditions to breed and continue the species: this is organised by man. The farmed salmon, because of its life under high population density, will often be diseased.

Some inevitably escape during transfer procedures or through damaged nets. These diseased 'escapees' may pass on to the wild salmon their infections and infestations and may also interbreed, producing offspring unsuited to the wild habitats. Furthermore, some farmed salmon are sterile. Certainly, the case that farmed salmon damage wild salmon is not proven. But the worry is very real, and many conservationists are deeply concerned about the expansion of the farmed salmon industry.

There are also concerns that the general damage to the environment caused by salmon farming will also damage wild salmon. Industrial pollution entering the rivers may be a factor in reducing salmon numbers, and if salmon farms were located adjacent to the river mouths, this would be expected to produce a particularly adverse effect.

Others believe that overfishing has been responsible for the decline in numbers. This could occur through unauthorised fishing in the rivers (or 'poaching'), or by excessive netting out at sea before the salmon return to their rivers. Certainly, new regulations have been introduced that are aimed to reduce widespread commercial poaching. But none of the problems seem essentially novel, and the question marks remain over farming as a cause of the demise of the wild salmon.

Fats of farmed salmon

The first clue that something may be amiss with the fats of farmed salmon is the observation that compared with the wild fish, the farmed variety is more likely to suffer from disease of the arteries and heart.

The most obvious difference between the fats of wild and farmed salmon is in the total fat present. That in farmed salmon is typically 4–6% of total weight, that in wild is 2.5–3%. The farmed fish does look more oily and the differences between the light and dark striations of the flesh seem more marked. This is analogous to applying an oil to dry wood to bring out the grain! The other changes concern the relative properties amongst the fatty acids. Farmed salmon has much less of the desirable n-3 and more of the less useful n-6 than wild salmon.

The question now is, should nutritionists enthuse over the desirability of farmed salmon as ideal food? It is not clear at all, and the realisation that farmed fish fats have changed for the worse has raised more questions over other intensively reared birds and mammals (see below).

Smoked salmon

Smoked salmon used to be the height of luxury; thin slices with brown bread and lemon. Wonderful. Now the glamour is going from farmed salmon. There is yet another problem. Smoked salmon is cold-smoked. A commonly used system is to burn oak-wood shavings with just enough air to keep the smouldering going but not to permit them to burst into flames. The smoke is carried through ducts to chambers holding trays of raw half-salmon cuts. The volatile and aromatic substances from the wood will permeate into the flesh of the salmon over the several hours' exposure. The salmon will become warm, but not hot, and the substances from the wood may not exert much effect on bacteria. You may be beginning to sense the problem: this type of cold-smoking results in a final product that is essentially raw and has been kept warm for hours.

Now, if meticulous care is taken over the whole process, then it may be safe, but the problem develops from the initial farming method, that is likely to encourage bacterial contamination in the first place. This may be initially confined to the outside surface of the fish, but de-gutting and halving the flesh are likely to spread any contaminating bacteria over most of the surfaces.

Two consequences of this have already occurred. First, the American FDA now regularly tests imported smoked salmon, and some batches from the UK have been rejected on account of the presence of listeria (see Chapter 9). Secondly, attempts have been made to outlaw the practice of sending smoked salmon by post in the UK.

It is interesting to note that there is no proof of ill health in any person resulting from eating smoked salmon. One wonders whether the discrete smoking of wild salmon would have provided such dangers, or indeed would have provoked such responses as have occurred to the mass production of a potentially dangerous food item.

In conclusion, the rapid expansion of intensive farming of salmon produces worries comparable to the problems associated with terrestrial systems: these include concerns of safety, the quality of nutrients of the food produced, and the high cost to the environment. It is true that farmed salmon on the whole looks and tastes like the genuine product and is of course cheap and is available throughout the year. But the price paid in other ways is very high.

Intensive rearing and polyunsaturated fats

The realisation that farming salmon does adversely affect the nature of its fats, raises questions over fats in intensively reared animals and poultry in general. I will first try and explain the types of fat present in food. They are generally present in two forms. One is the visible deposits of white or yellow material. These are made up of compounds of fatty acids and glycerol. The other type is invisible, and found in all vegetable and animal cells, with the fatty acid linked to proteins. Unprocessed lean meat necessarily contains a moderately high proportion of fat as do nuts and avocado pears.

Most fatty acids in the adipose tissue (and hence fatty meat) of ruminants – cattle and sheep – are saturated. Stearic acid is the most prevalent. This molecule contains a long line of carbon atoms and all the possible sites for hydrogen atoms to be attached to the carbon chain are occupied. It is said to be saturated, because no more hydrogen can be added. Its formula is $CH_3(CH_2)_{16}COOH$, simplified to $C_{18}H_{36}O_2$. This fatty acid tends to be solid at ordinary temperatures and is very insoluble in water.

Suppose the long carbon chain is two hydrogen atoms short.

This dramatically alters the properties of the fat, now a mono-unsaturated fat. The formula is now $C_{18}H_{34}O_2$, apparently very similar to stearic acid. But the new chemical compound has a kink in the molecule, it is an oil at room temperatures, and it mixes more easily with water. Indeed, the chemical is well known: it is oleic acid, the main ingredient of olive oil. Oleic acid is found in many (probably all) mammals' milk. Completely lean meat is also rich in oleic acid and some other mono-unsaturated fatty acids.

We lose two more hydrogen atoms, to make $C_{18}H_{32}O_2$ or linoleic acid, which is found in many vegetable oils such as sunflower oil. This is a polyunsaturated fat. To lose two more hydrogens we produce $C_{18}H_{30}O_2$ or linolenic acid, found in more vegetable oils, taking the name from linseed oil. This, too, is a polyunsaturated fat; as hydrogen is lost, so the molecule develops more kinks and the fat is easier for the body to use.

Indeed, a certain amount of polyunsaturated fats is essential. The problem is that animals do not make these to any extent; it is only plant life, either in the sea or on land, that produces these, through the action of light. So to obtain an adequate supply of polyunsaturated fats, we can either eat plant material directly or we can eat meat or fish that contains these as a result of their eating plant life. We have already seen that wild salmon have considerable amounts of their desirable n-3 polyunsaturated fatty acids.

The reason we are going into these details here is that the type of rearing of all sorts of food mammals and birds alters their type of fat, with a general deterioration in value of these essential nutrients as the rearing conditions become more intensive.

Cattle

Twenty-five years ago, Dr M. A. Crawford showed that the total weight of fat in wild buffalo (American bison, a ruminant) was 2.8% whereas that of beef in the butchers was 25%. Only 2.2% of the beef fat was polyunsaturated, compared with 20% of the buffalo fat. Moreover, the more active the buffalo, the greater was its proportion of polyunsaturated fatty acids to others (is this the scientific basis for jogging and sport as a whole?).

Efforts have been made to improve the amount of polyunsaturated fats produced in cattle. If cows are fed with supplements of soya oil, the amount of oleic acid in the milk increases. But attempts to alter the invisible fats in meats, by feeding cattle sunflower oil, have been less successful.

The main problem is that the animal's rumen has developed to make grass and other sources of cellulose digestible. There, bacteria break the big indigestible pieces of cellulose into small pieces which are then absorbed and used for various purposes in the animal's body. This also happens to sunflower oil, but the small pieces are not then reassembled into polyunsaturated fats. Many of the pieces are lost into the atmosphere as the gas methane.

One interesting finding from Japan is that when cattle were fed concentrates (mainly the protein-rich meal from rendering plants), whilst putting on more weight, and therefore of greater commercial value, there was a reduction in desirable fatty acids in the meat.

Pigs, broilers

With both pigs and chickens, the nature of the fats in the feed also has a clear effect on the type of fat in the resulting meat.

If pigs are fed supplements of sunflower oil, the fat in the meat becomes less saturated. Unfortunately, the more hydrogen atoms taken away from the fatty acid, the more it tends to be decomposed during storage, by oxygen in the air reacting at its weak links. This means that pork from pigs fed on oils can develop unpleasant smells or flavours. We will return to this general issue in Chapter 5, but suffice it to say now that the more complicated the food chain gets, and the longer the product has to survive in an acceptable form, the less desirable are polyunsaturates to the processor!

The same arguments apply very much to broilers, where the

composition of the animal's fat is influenced by its diet. Free-range chickens will tend to contain better fats than those intensively reared, for several reasons. One is that they have access to vegetation. The second is that their activity will tend to 'burn up' some of the saturated fat and thirdly, the temperature of the environment will be lower.

The real problem with broilers is that the artificial feeds associated with maximum weight gain are based on stored powdered cereals and protein concentrates. When fed these, the weight gain is remarkable, but the quality of the meat poor in terms of its fatty acids. The problem for the consumer is that salmon, poultry and red meat look very much the same regardless of the differences in the types of invisible fat.

The argument that it is preferable for us to eat directly the vegetable food, rather than give it to intensively reared animals, is beginning to make strong economical, nutritional and environmental sense; quite apart, that is, from grounds of animal welfare.

ADDENDA TO CHAPTER 4

1 I have just learnt from reliable sources of the following incident during the summer of 1992.

A salmon farm net enclosure in Loch Feochan, near Oban, Scotland, held about 20 tons of salmon; these were adult fish of both sexes, although many of the males would have been sterile, and so incapable of breeding. An accident occurred with a motor boat that resulted in serious damage to the net and 18 tons of salmon escaped. During the next few weeks, witnesses observed a much greater than expected migration of salmon up the adjacent rivers.

Comment

These 'escapees' must be expected to damage the wild salmon stock, for a number of reasons. They will compete for food, space and breeding grounds, and if they were to interbreed with the wild salmon, some of their genes would be likely to reduce the vigour of the resultant fish.

2 The EC is now proposing to ban canthaxanthin in the feed of mammals, birds and fish, but still permit it as an additive to foods directly! Zoo owners are now concerned that their flamingoes will not 'pink up' because of this ban!

PART II
Food processing

PART II
Food processing

5
Dried food

Introduction

Drying is the oldest means of storing food, making it usable during seasons or at times outside its normal availability.

Undoubtedly, man's ability to store food has been central to his global domination. The reason that removing water from food preserves it is that bacteria, the main cause of making food go off or bad (referred to as spoilage), do need adequate water to start to grow at all.

Drying will actually kill some bacteria, but often not reliably or completely. Many bacteria can change themselves into tiny hard spheres (technically called spores) as the amount of water present is reduced. The spores cannot produce disease in this form, but if the dried food is moistened and then kept wet, these spores can revert back to their earlier growing forms and be capable of either spoiling the food, or even causing illness.

This means that dried food will often contain many bacteria, but it is still safe, as long as it is eaten soon after rehydration. This is well illustrated with dried gravy powders that may contain as many as 100,000 bacteria per gram, but they are in a harmless, dormant state, and if eaten in this state, they are easily digested and destroyed by our enzymes in the gut. The danger comes from keeping the reconstituted gravy for too long and too warm. However, most bacteria take at least 2 hours to wake up from their dormant state and start growing, so one of the golden rules of food safety is that left-overs of gravies and similar products should be discarded after the meal, and not kept for the next. In practice this means that we should make up only the amount of gravy required for the meal concerned.

How dry must food become for it to be preserved? Surprisingly, quite a large amount of water can still be present in food that appears dry. Butter contains around 18% water, potato crisps (chips) and biscuits (cookies), say 5–15%.

Some food is dried naturally. The seeds ripening on plants contain little water; this enables the seed to survive in inclement environments prior to its deposit on suitably moist and fertile ground to germinate.

Nuts, cereals, and peas and beans (pulses) after harvesting are therefore inherently safe. The problems occur when the moisture in the air rises sufficiently for certain microorganisms to grow on the surfaces of these crops. The air contains tiny invisible particles of fungi, like bacterial spores, harmless to us in that state. But if the fungi alight on the moist outside coat of seeds, such as nuts, they can begin to grow, penetrating a network of minute and mainly invisible hairs into the seed. Fungi often grow in somewhat drier conditions than bacteria; one well-known problem concerns peanuts.

If peanuts are stored in warm, moist air, a fungus can grow and produce a chemical called aflatoxin. If these mouldy nuts are eaten, this aflatoxin can be absorbed into the body and cause illness. One effect of this is thought to be the risk of liver cancer.

This link has not been formally proved – and hopefully it never will need to be – because of the care and control over the movement and sale of such risky products. Dried foods such as nuts and dates are tested regularly for the presence of aflatoxin, and if the amount is above a certain level, the batch is destroyed.

The storage of animal feed powder under warm and moist conditions has also been found to favour aflatoxin formation, and one use of irradiation has been to remove this fungal contamination. Unfortunately, irradiation will not eliminate aflatoxin already in the feed, so this use, like the others, remains controversial (see Chapter 7).

There are a number of ways of drying food.

The simple exposure of food to the elements has been used from ancient times, and is still used for some fruit and meat. The tradition of exposing pieces of lamb to the air and wind is still to be seen on some of the Scottish islands.

Then some foods are inevitably dried as a result of heat used in the manufacturing process. Biscuits (cookies) and crisps (chips) are examples.

It has long been recognised that where food is dried slowly some problems do develop. Bacteria can have sufficient time to grow to produce toxins, nutrients such as vitamins and fats decompose, and the texture or feel of the food can also be spoilt. Some of the flavours and smells of food are due to substances that tend to be lost through evaporation during drying. And the slower the drying, the more these volatile chemicals tend to be lost.

So rapid and complete drying out of food has become a popular technique of the food industry and is usually performed through a process called freeze-drying. The food is first fast-frozen so that any water is rapidly converted to fine ice crystals. Then these are evaporated off under a vacuum and immediately the product is placed into jars, cans or packets that are sealed to ensure no penetration of air and its inevitable moisture.

Freeze-drying is used for instant coffee and tea, dried milk, gravy and soup powders. The storage life can be very long – up to several years – as long as the seal is kept intact. Most bacteria survive this process, but are obviously immobile and inert in the dry powder. This method certainly does hold some of the product's flavour better than other slower types of drying, but even so, flavour is lost and this accounts for the differences between powdered coffee and 'real' coffee.

However, with soups the flavour loss may be greater than with coffee; so the flavour enhancer monosodium glutamate and other artificial chemicals may be added to boost the taste. The result is that the final product will contain more salt than the original product and, of course, numerous chemical additives. Also, the drying process might cause loss of colours, and these may need to be replaced. Sometimes other chemicals are needed to prevent the powder forming into lumps.

It has been suggested that dried food is generally safe as regards bacterial contamination. This is not quite true: a few problems have occurred, but these must be rare compared with those of moist processed food. Batches of pasta have become contaminated with a bacterium called *Staphylococcus aureus*, which might cause problems after the pasta was rehydrated, and salmonella has very rarely been found in powdered milk. The problem here was that the bacteria were introduced into the food *before* it was dried, and subsequently survived.

Chocolate

Chocolate is one of the staple luxuries of the developed world. It is not too bad for you nutritionally, certainly not in moderation. It can be manipulated into endless varieties of shapes, additions and titillations. And it can be stored at room temperature! Cocoa forms its basis and itself makes a rich drink. The most luxurious sweets are made in Belgium, France and Switzerland. In other countries, the products tend to be more simple and cheaper. I find American and British chocolate a bit boring!

Where does chocolate come from? The story goes (and who am I to challenge it?) that the ancient Aztecs and Mayas (remember the Mexican pyramids) had been putting crushed cocoa beans into water to make a drink called *chocolatl*. It was usually drunk cold and all manner of spices were added. The 'magic' beans were taken back to Spain in 1502 by none other than the marauder Christopher Columbus. From here the drink became sweeter until it reached London in 1657, with the opening of the first chocolate house in Bishopsgate. Chocolate soon became popular among those who could afford it, and over the years the increasing consumption in Western countries has put pressures on increased production and reduced cost of the raw product cocoa.

It is just possible that in the last few years savoury snacks such as crisps (or chips) have knocked chocolate off its pedestal as the World's No. 1 snack, but accurate figures are not available. Perhaps we have had too much of a good thing?

The natural habitat of the cocoa tree is near the Equator, often not too distant from the coast. West Africa is the principal producing region of the world, followed by the West Indies and South America.

The production of cocoa is labour intensive. The initial tree seedlings require moisture and light, but not the direct sun. The cocoa crop does not appear for several years, and does not reach its maximum for up to 15 years. Oddly the flowers and subsequent fruit sprout directly from the trunk and branches. The cocoa seeds or beans are held in a large reddish-brown pod about 20 cm long and are picked by hand, usually twice each year. The picked pods are allowed to ferment in heaps to remove the pulp surrounding each bean. Then the beans are dried in the sun and can be transported to all parts of the world. When they arrive at their destination, they are cleaned, roasted and crushed, then the small pieces of edible chocolate (called nibs) are retained. To make chocolate, the nibs are ground into powder or cut into flakes, usually with the addition of milk products. The slurry is further processed according to the texture needed and poured into moulds to set.

The end result contains 2–10% water and should be too dry for bacteria to grow. It is not often realised that salmonella can persist in chocolate. There were several small outbreaks in the 1970s and 1980s in Canada and the USA. The key to understanding these accidents is that in the making of chocolate, the temperature cannot be too high as it results in the ugly whitening of the product. Exactly how salmonella becomes incorporated into the chocolate is not known, but it could well have been introduced through contamination of several ingredients by human handling.

However, the largest outbreak occurred in 1982 and 245 people in UK suffered from food poisoning due to a particular strain of salmonella, known as *Salmonella napoli* (so named after its Italian origin). The infection was traced by clever detective work to two types of chocolate bars, 'Tommy Junior' and 'Rocky Junior', both made in Italy. What was particularly interesting in this outbreak was the small numbers of salmonella bacteria actually in the chocolate, about 50–100 per bar; it had previously been thought that hundreds of thousands of salmonella bacteria were needed to cause an infection.

Presumably, the salmonellae in the chocolate were protected

by damage from the acid in the person's stomach after being eaten. They would then pass into the intestines and increase in numbers to produce bellyache and diarrhoea. That small numbers of salmonella in fatty food can cause disease is important in regard to the problem with eggs (see Chapter 9).

But these accidents should be avoidable, and dried foods must overall be safer than moist, preserved products. Enjoy chocolate as a luxury!

Biscuits, cookies

Biscuits and cookies are not quite the same. Although one is a British word, the other American, the British biscuit tends to be flatter and harder (and sometime duller to eat) than cookies which can have cake-like aspirations. The methods of manufacture are similar, and I hope American readers will forgive the use of the word biscuit here.

Chocolate, we have seen, is largely produced in the Third World and consumed in the developed countries. The food industry is central to this function and the problems resulting from eating moderate amounts of this luxury (yes, I believe it is a luxury) are few indeed.

Biscuits are in a somewhat different category. With a certain amount of care, most could be made in the home, although a few of the raw ingredients may be difficult to obtain. The commercial biscuit's success depends not so much on providing availability of a product that would otherwise be impossible to obtain, but on its convenience and also on its low cost. Making small numbers of biscuits in the home, whilst great fun for the amateur cook, is laborious and needs considerable fuel energy. The actual cost of the home-made biscuit may be as much, or even more than the retail brand. No wonder supermarket biscuits are so popular and seem to occupy a large part of the shelves! Incidentally, the occupied space is proportionately greater in the UK than in the USA.

Methods of biscuit manufacture are all fairly similar. The principle ingredient is a cereal, mainly wheat, maize, oats or rye.

According to the desired final texture this is ground either to a coarse or to a fine flour, water added, and the thick pastry-like slurry portioned and baked.

Crispbreads, apart from a small amount of salt, contain very little else and it is difficult to find any fault at all in most of these! But the sweeter and shorter-textured biscuits have substantial additional ingredients, principally fat and sugar. The fat is used for shortening – producing a firm but soft texture – and the sugar panders to our love of sweetness. Because fats, particularly polyunsaturated types, tend to decompose on prolonged storage to give a rancid flavour, the ones used often include hydrogenated oils and palm oil. It is these that have been producing doubts for nutritionists and will be discussed later (see Chapter 15).

Biscuits are essentially crisp-textured and their packaging and shelf life are aimed at ensuring that the crispness is retained. The packaging may not be completely airtight, because of cost, but if biscuits are stored in a dryish atmosphere – with a relative humidity of about 50% – absorption of water will be slow. Limp biscuits may be disappointing to bite, but they are probably quite safe!

Breakfast cereals

Breakfast cereals are one of the other successes of the food industry. Successful because a new form of food has been created that cannot be made by the consumer, except with extreme difficulty. Also, because the product is dry, it is safe. Each brand is based on a particular cereal crop that is usually ground, converted into a slurry, shaped and baked as with biscuit manufacture. Early this century saw the development of the classic types of cereal such as Kellogg's Cornflakes, Shredded and Puffed Wheat, Weetabix, All-bran, and Grapenuts. Most of these have stood the test of time and remain justifiably popular. Some contained added vitamins and minerals – not so much because of an awareness by the industry that its processes would destroy these, but more as a sales gimmick. In the early years, such vitamins might have sometimes been helpful in the then meagre diet, but today there really should not be a problem of vitamin intake – unless one 'lives' on beefburgers (perish the thought)!

Seen from a marketing point of view, the addition of a new brand would expand the total amount of breakfast cereals sold, presumably at the expense of bread, eggs and bacon. But ultimately the market had to become saturated, and the competition fought back. Remember the slogan in the UK, 'Go to work on an egg' just a few years ago.

So the pressure for novelty in cereals has mounted. Some were sold already sugared or frosted – to save the hard pressed breakfaster the unnecessary hassle of sprinkling the flakes with sugar! Then the packets contained the outlines of toys that could be assembled by cutting and glueing. Then actual toys were put into the cereal.

In order to promote the idea that the cereal was a complete food, almost everything was added: raisins, sultanas, honey, cocoa, and sugar – always sugar. The manufacturers of the early cereals had acted responsibly over the amount of sugar they added. This is not now always the case. Many are now too sweet.

Then the shapes changed. Recognising that not only did children have a sweet tooth, but were naturally aggressive, some of the shapes and toys became quasi-military, often with space exploration ideas. The number of types of cereal available increased further with the supermarkets selling their own brands, copying some of the standard products. You do not know whether these are in reality identical to, say, those made by Kellogg or Quaker, but in different packets, or are cheap imitations of lower quality.

The packaging has become more complex and costly, and I suspect that there is more cereal wasted because too many varieties are available and bought. The food industry tells us that the greater the choice, the better off we are. I am not sure about this: surely more choice in a fixed volume market means more cost in manufacture, greater need for storage space, and more stale food?

In conclusion, these old well-known brands are recommended: Kellogg's Cornflakes and All-bran, Weetabix and Shredded Wheat.

We will be returning to the role of cereals in the ideal diet in Chapter 15. But for the moment, we must recognise the value of the principle of the breakfast cereal. It is up to you to chose those which are both 'good' for you and that you enjoy. Certainly cereals are one of the major successes of the food industry. Well done, Kelloggs.

Crisps (or chips)

It is not exactly certain where and when the first commercial crisps were sold commercially. There is more than one claim for the original crisp and it is likely that initially there were many local products, some of which have been copied by national distributors.

Incidentally, we had better understand what we are describing. In the UK, packets of fried thin slices of potato are called crisps. In North America and many other countries, they are called chips. Of course, these products are not based just on potato, maize being commonly used.

The range available in a typical supermarket is now daunting. This was not always the case. The author's memory goes back to the post-war years when just one variety seemed to be available nationally! Smith's. The bags did not identify details of the contents, but were presumably slices of potato just fried and the salt – seemingly a *de rigueur* requirement – put in twists of blue paper.

Sometimes the bags were punctured and the crisps were soft. Over the years other brands began to compete and then audaciously, some manufacturers sprinkled the salt onto the crisps for you. This was soon to pave the way for sprinkling on artificial flavours and colours. Some types seemed to be short-lived, but others have endured, such as cheese and onion. Those without additional flavours came to be called plain and the exotic tastes of the others were highlighted by packet colours – beef was brown, tomato sauce red, cheese and onion yellow and green, and salt and vinegar blue or inexplicably, green.

From time to time reactions to all these different brands have set in, with the launching of real, traditional or original crisps, unflavoured but complete with salt in the blue paper twist.

All sorts of other developments have occurred, too. The potato has been dried, powdered, or reconstituted into slurry to make square crisps. The amount of fat used has been reduced (these are called low-fat, but you could argue that they should be called high-starch). Now, once the idea that potato crisps or chips could be fashioned from reconstituted potato powder, then doors opened to almost any shape, size, colour or composition. Some are made in the image of people or spaceships. The potential for variety is a bottomless pit.

Then, for the calorie-conscious public, some are aerated so the product is full of healthy bubbles, as are some brands of chocolate.

Many types of this packeted snack genre are not based on potato at all. Maize is popular. Or any old starch – re-fashioned, blown up, dyed, flavoured and sold with all the hype of mass media advertising and packaging.

The supermarket crisp shelves are a veritable Aladdin's cave: a paradise for children. Health freaks can choose types that are low-fat, calorie-reduced, or high-fibre – potato skins or dried apples. Indeed, the boundaries between crisps and nuts and dried fruit have become blurred. Anything goes. The dried packeted snack rules the Western World!

Well, why not? Is there anything wrong with crisps or chips? Not when eaten in moderation, but there are some concerns.

The first is that crisps usually contain around 30% fat. This may not be a problem, depending on the type of fat. Take slices of potato fried in sunflower oil, packeted and soon sold. No worries here. But suppose these crisps do not sell rapidly, and after a week or so the sunflower oil starts going off – rancid. Then the crisps will taste and smell 'off' or stale.

As with cereals, the more varieties there are available for a fixed market volume, then the trend must surely be for each to be stored for longer. So the combination of the ability to manufacture such a variety of snacks, and the hunger of the consumer for them, make the problems of storage very real, space apart. This is why more and more of certain types of oils are being used with the intention that they will enable the product to last for longer before being eaten.

It is unfortunate that polyunsaturated fats do tend to decompose more readily than saturated, so the change in the oil

is frequently towards types which are either saturated or chemically treated by hydrogenation to make them last longer. Palm oil is an example of a saturated vegetable oil and is obtained from the coconut palm and the main fat present has the formula $C_{16}H_{32}O_2$ – known as palmitic acid. You see the ratio of hydrogen to carbon, like the principal ingredient in mutton fat, is exactly 2:1.

Hydrogenated oils

One way to stop polyunsaturated fats from going rancid is to treat them with hydrogen. The end result is known as a hydrogenated oil and many varieties of processed food will list this as an ingredient. It will rarely be listed as an exact amount, as it will be difficult to know exactly how much of the oil has been altered chemically. Hydrogenation is used by the food industry in two ways: one is to make the oil last longer before going 'off' and the other is to solidify fluid oil. You may wonder how pure fluid sunflower oil is converted to solid margarine made from 'pure sunflower oil'. The answer is through hydrogenation.

It is interesting that the use of hydrogenation was developed by the food industry with little research or concern for its effects on us, the consumers. Is there a problem with hydrogenation? Yes, there may well be, because the resulting fats produced are often unnatural. Chemically, the shape of natural unsaturated fats is such that they are said to be in the *cis* formation. You may have seen that the main component of sunflower oil is *cis,cis* linoleic, meaning this fat has two regions of its chemical molecule in the natural shape.

When polyunsaturated fatty acids are hydrogenated, about half the resulting chemicals are in the unnatural or *trans* shape. This may well matter a great deal because experiments have shown that the human body does not seem to handle this unnatural fat as well as the natural. Experiments in Holland were performed on 34 women and 25 men who volunteered to test different diets. Each received identical diets except that 10% of it was varied in the fat supplement. Some received a natural *cis* unsaturated fat supplement, others the unnatural *trans* fat and others the saturated fat palmitic acid. Each person, in variable order, ate each of the diet types for 3 weeks, so the experiment lasted 9 weeks.

The effect of these diets on the subjects' blood cholesterol was assessed by taking samples of blood at the beginning, and at the end of each of the 3 week periods. With both the saturated fat, palmitic, and the unnatural *trans*, the blood cholesterol rose. Furthermore, other blood changes occurred when on the unnatural *trans* diet, that were thought to be 'unfavourable'.

These studies suggest, therefore, that people who consume substantial amounts of hydrogenated oils, like saturated fats, are more likely to develop raised cholesterol levels in their blood and so possibly be at greater risk of heart attacks and strokes.

Further studies are urgently needed in this area and we should look at more people for longer, in different countries and on different types of diet. As we will be discussing in Chapter 15, the value of levels of cholesterol in the blood in predicting heart attacks is not simple, although I think there is complete agreement that our diet should not make it go up too much.

But surely, the regulatory authorities should, at least until evidence is established to the contrary, view hydrogenated oils as artificial additives and treat them with the same scrutiny as, say, cyclamates or saccharin?

It is a reasonable approach now to try and eat only dry fat foods, such as biscuits or crisps, that contain vegetable oils other than palm oil, and not hydrogenated oil. If such a product has a pleasant taste it is likely to be fresh in the true sense of the word.

Recommended dried food on the basis of safety and nutrition*

Soups

Gravies

Custards

Coffee and tea

Pasta

Chocolate (occasionally)

Cookies and biscuits not containing hydrogenated or palm oil (occasionally)

Chips (crisps) not containing hydrogenated or palm oil or too salty

Breakfast cereals that are not too sweet

* These products will be low in certain vitamins, some will be low in protein and fibre, and should not be viewed as providing all or even most of the dietary needs.

6
Bread food

Introduction

We have become so familiar with bread as a food eaten and even sometimes enjoyed daily, that we forget that it is not really part of our natural diet in the biological sense, and that it was one of the first products that we learned to process. We do not actually need to eat bread at all, but the best examples are highly nutritious and thoroughly enjoyable. The worst is a routine eating chore performed out of convenience, and of poor food value. We will be looking at the great bread debate later, but first consider the origins of bread.

When ancient man first started to grow wheat and other crops for food, he must have already been aware that he could not eat the grains unless they were first ground up. If you don't believe me, try eating whole wheat or maize grains and they will probably pass straight through your intestines, undigested. This is because the tough outside coverings of the seeds prevent our digestive enzymes from access to the nutrients underneath.

Our teeth, we saw in Chapter 1, evolved mainly to remove the outsides of large fruits, not to crunch whole grains. (Don't try to do this, as your teeth may break.) But birds, such as chickens, have evolved the ability to digest grain. They possess an instinct that makes them peck sharp grit which roughens the outside of the grain sufficiently for their digestive juices to get into the inside and dissolve the nutrients so that they can be absorbed.

Whilst bread or other processed cereals may not be our natural diet, we as a civilisation are now completely dependent on them and we must make the best of them. Without cereals we would be really stuck for sufficient food.

Bread, cakes and similar food contain moderate amounts of water, and are usually described as 'intermediate' in water content. The amount of water is more than in dried food, but less than in wet food, such as meat. This amount of water in bread and cakes is critical since it is adequate to permit fungi and yeasts to grow, so enabling these products to be made to rise from gases released by growing yeast. But the water is not present in sufficient amounts to allow most bacteria to grow. This means that bread and cakes can be stored safely at room temperature. If there is a problem from too long storage, it is due to drying out that makes the product 'stale'. Stale bread is, however, safe to eat, even if not as enjoyable as fresh.

Exactly when bread making began is not known; perhaps primitive *Homo sapiens* found that if he used stones to crush acorns from the oak trees in the deciduous forests, and added water, he could produce a tasty 'scone'.

It seems that initially the grains of cereals were crushed between two smallish stones, manually. But in ancient Egypt, the milling, as the crushing has become called, was scaled up, with the millstones rotated by oxen. The whole grain would be used and the particle size of the flour produced would be variable.

The addition of water, some seasoning and heating produced unleavened bread. But the addition of yeasts to lighten the texture has been used for thousands of years.

Whilst some yeasts can cause illness, those used in food do not, because they have, through the centuries, been adapted for the various functions in food. Bread yeast is different from brewer's yeast, for example.

The bread yeast grows without very much water, and has the remarkable property of speed of action. Makers of homemade bread can see the dough begin to rise after 30 minutes, or even less!

The gas produced by the yeast is mainly carbon dioxide, released from the carbohydrate and protein present. Other substances are also formed, which helps to explain some of the glorious tastes and smells of newly baked bread. Even traces of alcohol occur, and of course brewer's yeast has been specially developed for that purpose. (Before teetotallers abandon eating bread, traces of alcohol are also made by our own bacteria in our intestines!)

In cake manufacture, in contrast to bread, another raising agent

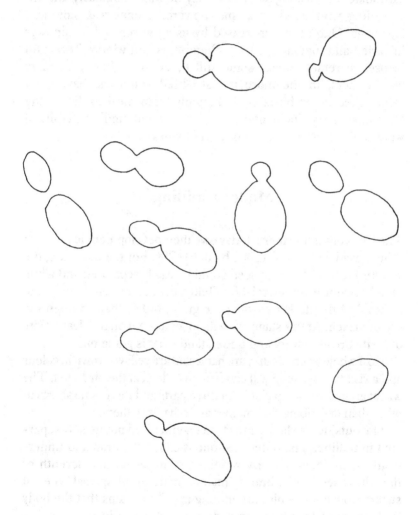

Budding yeast

has been used for centuries: this is bicarbonate of soda. When this chemical is heated, carbon dioxide is released and sodium carbonate remains. This has a soapy or metallic taste, which is an immediate give-away if manufacturers have taken short cuts in getting bread products to rise. Indeed, some of the artificial additives found in cakes are used to disguise the taste of sodium carbonate, or washing soda as it may be more familiarly known.

Milling with hand or animal-powered stones was slow and laborious. Output was increased by using water and windpower in varied and ingenious forms. Watermills and windmills remain in use: in remote areas, some still function in their customary way. Others, in the more industrialised countries, have been refurbished by enthusiasts, and supply fresh meal to discerning local customers. Incidentally, most of the windmills in Holland were built to drain water, not to mill wheat.

Modern milling

The types of milling in use today owe their development to the steel rollers used widely in Europe by the 1870s. For the first time, the various parts of the grain seed could be easily separated and white bread became a commercial feasibility, the colour being due to the removal of the darker parts of the grain, rather than through the use of bleach. At this stage, we will not comment on the desirability of white bread, but try and understand what is going on.

Most whole grains as they are harvested are yellow-brown in colour and a visit to a pet shop will confirm this – look at the bird seed. The wheat grain is made up chiefly of three parts and there is no short cut other than describing their names and what is in them.

The outside two layers are close together and not usually separated in milling. The outermost one is called the bran and underneath it, the aleurone. By weight they make up one seventh of the wheat seed. The bran is mainly made up of special types of sugars whose molecules are in long invisible chains that the body finds difficult to digest. Bran does not dissolve in water and is usually referred to as fibre or roughage. There are also present

Vitamins E, B_1, B_2, B_3, B_6 and folic acid, and the minerals calcium, magnesium, potassium, and iron.

The second region of the wheat seed is called the germ. It is from this that the new seedling plant emerges when conditions are right, and the process is, of course, called germination.

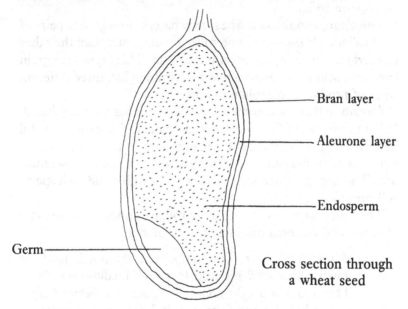

Bran layer

Aleurone layer

Endosperm

Germ

Cross section through
a wheat seed

The germ is only small – about one fortieth of the total weight of the seed – but packed with nutrients – almost everything needed for life: unsaturated fats, proteins, vitamins and minerals.

The rest of the seed provides the store of nutrients for the germinating seedling, accounts for over four fifths by weight and is known as the endosperm. It does contain small amounts of B vitamins, and traces of minerals, but is mainly starch and the protein, gluten.

The modern milling process – with around 25 sets of steel rollers – generally occurs as follows. The wheat is sometimes transferred by lorry directly from the combine harvester to the factory. Frequently it will be stored first for, hopefully, not longer than a year or so, because beyond that time deterioration sets in. During storage, the grain may well be contaminated by dust and parasites.

So the first job in the factory is cleaning away any extraneous matter such as grain husks, other seeds, pieces of straw, stones and so on. For the successful separation of the various parts of the seed, the water content of the wheat is crucial. So in effect, it is air-conditioned to make the bran tougher and the white endosperm softer.

The clean, conditioned wheat now passes through five pairs of steel rollers. Of each pair, one roller rotates faster than the other so exerting a shearing force onto the grain. This breaks the grain open and with subsequent filters and air draughts, three different types of product accumulate into heaps.

Already at this stage, some pure white flour has been produced. Other products are fairly coarse white pieces with some bran still intermixed. The bran is lighter than the white endosperm and is removed by air pressure. This material is described now as semolina. The final product is rich in bran, but with some endosperm still present.

The process continues with more rolling, grinding, sieving and blowing, and the final products are as follows.

1 **Wholemeal flour**. This contains all the fractions of the initial wheat seed reduced to a fine–medium powder. The colour is a light grey-brown, actually lighter than that of bread made from it. It is 100% pure wheat: nothing added, nothing extracted. One of the leading protagonists of this product is Allinsons in the UK, named after Dr Allinson, with their advertising slogan 'Bread wi' nowt taken out'. So wholemeal flour generally has the greatest content of minerals, vitamins and fibre (bran).

2 **Wheatmeal flour**. This product has been called brown flour and contains about 85% of the wheat seed – that is, most except the outermost layers. This product will be missing much of the fibre of wholemeal flour and some of its vitamins and minerals. Some of the brownness of brown bread is actually faked due to the addition of caramel. Caroline Walker, in her book *The Food Scandal* (co-author Geoffrey Cannon) commented

that such faked bread can sometimes be detected by
tell-tale swirls of white in the finished loaf.
3 **White flour**. This is just the ground-up endosperm,
full of protein and starch and not much else. It accounts
for about 72% of the total weight of the wheat grain.

There are variations on these flours. For example, where exclusively wholemeal flour is required, there is no need for such a
complex milling procedure. Indeed, Allinson's are proud of their
old-fashioned, stoneground method of milling. Most of the other
cereals tend to be milled in their entirety and may be obtained
on their own or mixed with other flours, such as wheat.

Making bread

Makers of bread in the home will be familiar with the way it is
made. For others, this description is given to show how easily
the task can be undertaken, even by a man!

There is a great deal of mythology over temperatures, times,
and special ingredients or mixing devices. For me the only absolute need is a place sufficiently warm – say, at 20–28 °C – for
the dough to rise. On occasion the greenhouse has been used!

To the flour are added water, salt and yeast, and you can start
the yeast growing separately if you like. The mixture is sticky and
elastic and tends to cling to your fingers. You might find it on
tap handles and door knobs days later!

The reason that the dough is elastic is because of the presence
of the gluten protein, and kneading produces a network of gluten
in the dough; in between the network of gluten, the yeast as it
grows releases the gas carbon dioxide which becomes trapped so
the dough rises.

The second kneading is used to ensure the uniform distribution
of the gas. The finally risen (hopefully) bread is baked in a hot
oven to kill the yeast and precipitate or fix the protein–starch
network. The final product might appear fairly dry but will contain
about 25% water.

The key for bread to rise is the presence of the network of gluten/starch; that is derived from the inside of the grain, or the endosperm. The presence of the bran is not involved in the process of fermentation by the yeast, so it is literally a dead weight on the yeast. Herein lies the whole dilemma with bread: wholemeal flour rises less well than white flour. White bread is therefore lighter (or less dense) than wholemeal and many types of brown bread. We will return to this issue in due course. But for the moment, the home baker is only too aware that attempts to make bread without at least some white flour may produce virtually unleavened results. Pure rye or barley flour produces something more akin to biscuits than bread.

There has been some confusion over what is meant by strong or weak, and hard and soft as applied to flour and bread. I'll try to explain. Some wheat seeds (e.g. Canadian) have a high protein content, so these generate **hard flour** because more energy is needed to convert them to fine grains of flour than, say, lower-protein European types (known as soft). A **strong flour** refers to the type of dough that is produced, with strong meaning a dough that resists being stretched and weak meaning one that yields readily on being pulled.

Bread is usually made from strong flour, and biscuits from weak. So for the amateur breadmaker, I suggest you begin with strong white flour and develop it from there, once you have established your own conditions that produce the sort of product you want. Quality light breads made with wholemeal flour are a real challenge, particularly if you don't cheat by adding any white flour, or even a suggestion of baking soda!

One of my favourite breads is potato bread, made by replacing some of the flour by mashed potato and reducing the water. Making bread is fun, and fun is something that seems to have disappeared from so much of our eating. Each batch of loaves produces its own unpredictability and excitement. The smells are delicious and the real treat is to eat it when warm. Most home-made bread recipes do not result in a product that stores well for too long, so do not make too much at a time, and tend to make it on the dry side.

Commercial bread

The procedures used for commercial bread (i.e. bread in the shops) are in general similar to those employed for making bread in the home, the main variable in both being the composition of the flour.

White or brown?

Ever since it was possible to produce white as well as wholemeal flour, the arguments have raged as to whether we should be eating white or brown bread. What is good for us is only part of the issue; the others are what we want, what we really enjoy and what we can afford. The issues are complicated and I am indebted to Geoffrey Cannon's publications for some of the material.

We have already seen that bread made from white flour has fewer vitamins, minerals and fibre compared with wholemeal bread, although today's white bread has some of these replaced. This does not automatically prove that white bread is inferior to brown, because if we are able to obtain these missing nutrients from other foods, then there should be no problem.

But instinctively, surely we must question the point of milling wheat in such a complicated way to separate all its components and then for us to eat just one. Would it not be more sensible, and surely in the long run cheaper, to mill to wholemeal only?

In the early part of this century, there were enough data to suggest that if children were fed white bread, they were more likely to experience nutritional problems than if fed wholemeal. But we must remember than in those days bread really was the staple food and this must have meant that the diets of the time gave an inadequate opportunity to make up those deficiencies in white bread. A League of Nations report in 1936 stated that

White flour in the process of milling is deprived of important
nutritive elements. Its use should be decreased and partial
substitution by lightly milled cereals . . . is recommended.

This had little impact and the consumption of white bread
remained high. Cannon has suggested that the main reason for
the popularity of white bread was that the millers were promoting
it for the following reasons: white flour keeps longer; it can hold
more air and water, and the unused bran is sold to the animal
feed industry, so the total volume of flour produced is increased;
lastly, wholegrain bread requires more skill and time to make.

Most of the desirable properties of white bread, as seen from
the point of view of the producers, are determined by the ability
of yeast to produce more retained gas in white flour than in
wholemeal. The lighter texture allows more water and so
improves keeping qualities, so essential for central manufacture
followed by distribution to a national network of retailers.

Yes, white bread does have some advantages for the manufac-
turer, but I do not believe these are overwhelming. After all,
Allinsons (remember, 'Bread wi' nowt taken out') do successfully
achieve wide distribution of their wholemeal loaves. I think, also,
that no-one will impose a product on the consumer that he or
she does not want. So this leads me to suggest that many consumers actually want white bread, for some of these reasons.
First, the colour: whiteness is symbolic of purity; grey is not.
Secondly, the texture is lighter and the bread tends to make
crisper toast than wheatmeal and absorbs more egg and so on in,
say, French toast and bread-and-butter pudding. Then, it does
keep well – at least the type that is mainly produced in the UK.

So to many people in the 1930s, white bread *was* bread. But
the arguments continued and following the declaration of war in
1939 in the UK, the national loaf – brown in colour and using
85% of the wheat grain rather than the 72% in white – was
introduced.

Whilst there was seeming agreement amongst nutritionists,
politicians and doctors that the national brown loaf was an excellent development, one wonders what the consumers thought of
it. Did they not associate the dreary colour, standard size and

inevitable tendency to staleness with the days of suffering and shortages of war? Were not the public yearning for their white bread?

They were soon to get it. The argument that white bread was deficient in certain vitamins and minerals was deflected to an extent by adding these to the white flour, beginning with Vitamin B_1. The evidence for the beneficial effect of fibre had not accrued, and a key experiment was performed in 1947 and 1948.

German orphans from Wuppertal, aged between 3 and 15, were studied for the effects of various diets on their development and well-being. One important factor in the experiment was that all the children were allowed to eat as much as they wanted, and any changes in the diet were within the relative composition of this *ad libitum* diet. In particular, all the children were fed lots of vegetable soup, and were also given vitamin and mineral supplements. This is very important, because when assessing the different types of bread and their nutrients, the findings must be complicated by the effects of the other foods eaten. (Incidentally, before you start criticising this experiment, it is obviously not ethical deliberately to put children on diets near to starvation. This has been the over-riding ethical dilemma for so many studies on nutrition.)

In the Wuppertal experiment, the main variations were in the type of bread: some were given wholemeal, others brown, others white and yet others white containing additional B vitamins, calcium and iron. The experiments lasted a year, quite a long time for such an investigation to be managed, but far too short a time for any really meaningful effects on development and disease to be detected. This again shows how extremely difficult it is to perform experiments that really do answer the important questions. Anyway, all the children flourished, and on the face of it, white bread was as nutritious as wholemeal. But the trial was flawed scientifically, in that it did not look at bread in isolation from other food.

These results were published by the UK Medical Research Council in 1954, by the main analysts of the trial, Dr. Elsie Widdowson and Professor Robert McCance. But the arguments continued and these authors wrote a book *Breads, White and Brown*. The chief conclusion was:

> Unenriched white flour is *likely* to be as valuable a part of the
> diets currently used in Europe and America as an enriched
> flour, an 85% or 100% wheaten meal . . . There is certain
> to be resistance among the health minded to any evidence
> that white bread might be as nutritious as brown.

Today, we are not really much further on: essentially the position is that white bread within a generally mixed diet cannot be
claimed to be harmful; in the long term the absence of some
substances – such as Vitamin E, found mainly in the bran layer
of the grain – might cause cancer in some people. This will be
discussed further in Chapter 15.

Over the years, the importance of bread made from wheat, in
man's diet in many developed countries, has declined. The Lord's
Prayer illustrates the vital role of bread in early times by the
phrase, 'Give us our daily bread'. So the function of bread has
changed, from being the essential central component of our diet,
to one that tends to be an appendage; to provide a means to eat
something else that matters. The sandwich illustrates this well.
Bread is needed as a means of holding the moist filling without
soiling our hands. At sit-down meals (amazingly, this category of
eating now has to be identified specifically) bread is reduced in
status to the side plate.

The consumption of bread has approximately halved in Europe
and North America over the last 40 years. It is no wonder that
these areas now produce a surplus of grain; of course, greater
yields are partly responsible.

I believe that one of the factors responsible for the decline of
bread eating is the extremely poor quality of many of the products;
I refer in particular to the centrally produced, white sliced and
wrapped loaf, that is moist and aerated to produce a bigger loaf
for the same ingredients that go into a traditional loaf. But it is
predictably boring in uniform texture and lack of taste and has
to be toasted, spread or somehow titillated to make it likeable.
However, it is cheap and available from many retailers.

Let us end this section on a positive note and make some
recommendations. No: I am not going to advise a boycott of all
white bread. What I propose is that among freshly baked local

breads (and the most local venue is your home!), wholemeal bread is one of the best and most enjoyable foods there is. It is probably a must where your food intake is only just adequate and also for people who do not eat meat or fish, and in particular for vegans who also avoid eggs and dairy products. It must make common sense for the whole wheat grain to be milled, and avoid the complexity of removing parts of the seed for making white bread.

But what about those of us who eat a varied diet; should we eat white bread all? Yes, of course, but please let us go for quality. The French loaf or baguette is an example of high quality food. The shape, whilst technically being inefficient as far as mass production goes, produces a large area of crust. The texture is light and airy, and it has to be eaten when freshly baked because of its low water content. It goes stale and dry after 24 hours. This means that it often has to be made locally, and so much the better. There has been a gratifying increase recently in local bread, including English versions of French bread produced actually on the premises of local retailers in the UK.

Surely this trend will continue; and let us work for the demise of the centrally produced, wet white sliced (or unsliced) wrapped loaf.

Pizzas

The Italian plural of the word 'pizza' is pizze, but not many people are familiar with this, so we'll stick to pizzas. Not just the name of the product has been debased, but so has its quality and enjoyment.

Pizzas come into this section, because more and more they are being made with bread-like bases. The traditional Italian pizza is made fresh with essentially flour and water, the top coated and covered with cheese, tomatoes and other local produce and then baked briefly in a very hot oven – around 650 °F (343 °C).

The crusting and rising of the dough base is achieved largely by fierce heat. Such pizzas are available in certain restaurants in North America and the UK but the degree of heat is difficult to achieve in the home, and so often the pizza retailed can only be described as a fake.

You may have noticed the tendency for 'deep pan' and other bread bases, often allowing the customer to select his choice of toppings. This has paved the way for an extensive market in pre-cooked bases, usually of a bread-like consistency, sometimes packed in vacuum packs. Indeed, there is one UK major retailer obtaining its pizza bases from the USA, claiming that it cannot manufacture them in the UK. Amazing.

Some pizzas are openly types of bread. Pieces of stale, white long rolls smeared with tomato purée and grated cheese are described as 'French bread pizzas' and can be purchased deep-frozen. Pizzas therefore can be presented in almost any way, the base can be fresh or reheated, the topping can be fresh or reheated, the whole item can be pre-cooked, transported, reheated, transported and reheated almost *ad infinitum*. The pizza as a food has become grossly devalued.

It is still possible to eat the genuine article outside Italy. One clue is the restaurant staffed by many people, and you should see the bases being rolled or hand-stretched. Be most suspicious of the solo cooks whose function seems to be to transfer frozen or chilled ready-made pizzas to ovens for reheating.

Other pasta

As with pizzas, pasta is cereal based. Most is made from hard or durum wheat that is milled into semolina and water, sometimes egg yolk, and vegetables such as spinach, added before drying into various shapes. I have never understood the fascination for the unwieldy length of spaghetti and sense is prevailing with shapes including short tubes (i.e. macaroni), shells, butterflies, stars, numbers, sheets (lasagne) and many more!

Pasta provides a good range of nutrients, and once dried, it can be stored for long periods at room temperatures. Egg is not actually needed in pasta production, and inevitably there is a flourishing market in cholesterol-free (i.e. egg yolk free) for the cholesterol-phobic American consumers.

Like the pizza, the problems can come when the catering industry takes short cuts: I know one chain of restaurants that makes up its spaghetti once a week and keeps it chilled in bulk pending an order, when it is reheated in the microwave. Is this edible food?

Recommended bread products

    Wholemeal loaves, freshly and locally produced, uncut

    French white bread, within minutes or a few hours of baking

    Rye bread

    Corn and potato bread

    Locally prepared pizzas

    Some breakfast muffins (instead of eggs)

7
Moist processed food

When discussing dry and intermediate processed foods in Chapters 5 and 6, some enthusiasm, at least for some products, was expressed because those systems have been well researched, have been in use for centuries, and are generally safe – certainly as far as bacterial contamination is concerned.

But when we look at moist processed and stored food, the potential problems are much more worrying. The food industry has addressed many of these, but not all, and despite careful monitoring, errors do occur from time to time. There are other, real problems: the storage of food with a high water content can result in loss of vitamins, particularly Vitamin C and folic acid, and there is an increased tendency for fats to become rancid. Moreover, the colours in the food can fade; the usual reaction to this by the industry is to add artificial colourants or other chemicals to counter these changes. People simply won't eat food that is the 'wrong' colour unless it is a cult product, such as canned yellow mushy peas!

There are also concerns about changes in proteins and sugars during wet storage, but very little research has been undertaken on these.

The problem bacteria

We need to know something about bacteria that grow on or in food in order to understand the problems. Take a freshly picked apple with an intact skin and no damage at the entry of the stalk.

There will be a few bacteria on the skin. These will have come from the picker's fingers, the air, the wrapping, dust and so on. They will tend to be dormant, being short of water and unable to use the nutrients inside the apple.

These bacteria will be eaten by you and will usually be digested and eliminated completely. A few might start to grow in your intestine, but these will be most unlikely to produce disease. Such bacteria will probably have come from the picker's fingers: these in turn may have come from his intestines. As with dried food, we do eat bacteria from fresh food, but on the whole they are harmless.

But suppose we store the apple. After a few weeks it might go mouldy and bad. This can be due to fungi that are in the air, which are able to absorb water and gases from the air that enable them to start growing on the apple skin.

If the skin is punctured – by, for example, the picker's finger nails – fungi or bacteria may start growing quite readily and the apple will go bad quickly. In general, as fresh food is stored it loses some of its natural ability to resist bacteria and fungi, so the risk of badness increases with age. The same applies to potatoes. Gardeners know that if they store, say, 100 potatoes over the winter, typically in the first month one might rot, in the second 3, in the third 6, in the fourth 10, in the fifth 20 and so on.

But suppose we peel the apple and chop it up and store it at room temperature. It may immediately become brown-coloured as a result of oxygen causing chemical changes, but soon bacteria and fungi will start growing, and produce unpleasant tastes and smells after 1–2 days and perhaps obvious growth of fungi after a week. If we cook the chopped peeled apple first and then leave it exposed, the decomposition will set in even sooner.

These changes in the apple also occur more generally with fruit, vegetables, meat and fish. The outer skin is an important barrier to bacteria and fungi, and processing – chopping or mincing in particular – and previous cooking all tend to encourage the growth of bacteria. Undoubtedly fresh is best, and moist food processing virtually never improves the quality, safety or nutritional value of food. But it can change the types of product and make them available during all seasons.

Bacteria in food are certainly not always harmful: many can be beneficial. When bacteria grow on the surface of meat, we can identify different types. Red meat of recently slaughtered animals is tenderised during hanging, in part by bacteria growing on the surface of the meat. These harmless bacteria are transferred to the meat from the animal's organs during slaughter, from the environment or from handling.

Other bacteria may be harmless initially, but after days or weeks, their numbers may build up to such an extent that they produce unpleasant smells, tastes (although you would not risk eating it, usually) and colours. For example, the green colour of bad meat is partly due to the presence of a bacterium called pseudomonas. This is one of a group of bacteria that are called food spoilage bacteria, producing nasty smells but not real danger. But they alert you to the possibilities of other bacteria that might be dangerous, which is why you don't eat the meat. It is 'off', or bad. These other bacteria usually have no taste or smell and are the food poisoning bacteria.

Suppose a piece of whole meat – say, a joint of pork – does contain some food poisoning bacteria, they will be found mainly on its surface. Cooking should readily destroy these bacteria because they are confined to the surface of meat. The colour change, from red to brown, on the surface of the meat on cooking is usually a good indicator that any potentially dangerous bacteria have been destroyed.

But if that joint of pork is minced and fashioned into pies, sausages, and so on, the bacteria on the meat's surface are distributed throughout the product, and there must be a greater chance of the sausage containing some bacteria in its centre after cooking than of whole unminced meat. These bacteria could include the types capable of causing illness.

Moist food processing must generate additional risks, and in this chapter we will consider some of the successful products where such risks are usually minimised, and other products which still seem to be something of a gamble.

Deep-frozen food

One of the most successful methods of preserving food is through deep-freezing. The usual temperature is between −18 and −23 °C. At this temperature all bacteria and fungi stop growing completely, but they are not often killed.

Deep-freezing also arrests, either completely or very nearly, the action of the enzymes responsible for decomposition of stored food. If deep-frozen food is stored for too long it may lose colour, texture and some nutrients but it should not cause food poisoning.

A good tip for freezing items in the home is to choose items as small as possible, and not to insert too many items simultaneously into the freezer. This is to encourage rapid freezing, when the ice crystals formed are tiny and do little damage to the product, whereas the larger crystals formed in slow freezing can be quite destructive.

Here are some housekeeping comments on the home deep-freeze. Cabinet types seem preferable to cupboard or combined vertical fridge-freezers because with these, on opening the door cold air literally tumbles out, so increasing the running costs. If models do not contain a thermometer, buy one: the temperature should be between −18 and −23 °C. If there is a power cut, do not open the lid until the power is restored. Then do so, and if the temperature is −5 °C, or lower on the top, shut the lid and all should be well. If it is above −5 °C, you may have to discard some or all of the items, keeping only those which are firmly frozen.

The star rating on frozen packets is only a guide, and applies mainly to quality, not safety. A common code is as follows: one star suggests a life of just 1 week, two stars 1 month, and three stars 3 months.

Many freezers undergo automatic defrost cycles. For those that do not, I suggest that a useful plan would be to clear the freezer of all food (planned, of course) every three months, defrost, clean and restock.

All items should be individually wrapped and you should have areas dedicated for either cooked and ready to eat or raw food. For example, on no account permit a naked ice cream to touch raw chicken! Ideally, items should be labelled and rotated to ensure first in, first out rather than the common practice of last in, first out! This is easier to state in writing than to achieve in practice.

Thawing

Not only are the small items preferred for the freezing process, but also for either thawing or cooking direct from frozen.

If there is a serious snag with the deep-freeze, it is with thawing of larger items. It is essential that any item be given sufficient time for thorough thawing before cooking. Raw poultry should be carefully contained during thawing so that any emerging juices (water is often added deliberately to increase the weight of frozen food, *oh dear*) do not splash onto other food. This is best achieved in the bottom of a refrigerator. Typical times are given in Table 1.

Table 1. *Guide to thawing poultry*

Method	Size of poultry	Appropriate length of time needed
Refrigerator	1–3 lb, small chickens, pieces	1 day
	3–6 lb, large chickens, ducks, small turkeys	2 days
	6–12 lb, large turkeys	3 days
Cool room	1–3 lb, small chickens, pieces	12 hours
	3–6 lb, large chickens, ducks, small turkeys	1 day
	6–12 lb, large turkeys	2 days
Microwave (read instructions)	1–3 lb, small chickens, pieces	8–15 minutes[a] (standing time 10 minutes)
	3–6 lb, large chickens, ducks, small turkeys	15–30 minutes[a] (standing time 20 minutes)

[a] Approximate, please read instructions

Cook–chill

The phrase cook–chill is a new one, which we could happily live without. Food is cooked in a factory, rapidly chilled (not frozen) and then transported, stored, transported and then finally sold, reheated and eaten. Some products are eaten without being reheated. The system has been deployed more widely in the UK than in the USA.

In 1980, the British government claimed that in institutions, cook–chill could be made safe, as long as the food was kept at temperatures between 0 and +3 °C, and the refrigerated food should not be kept for longer than 3 clear days between the day of production and the day of reheating and eating.

However, it was not until 1988 that the problem of *Listeria monocytogenes* in cook–chill food became well known. This bacterium has been found in anything between 2 and 50% of cook–chill food, is relatively heat-resistant, grows at temperatures down to 0 °C, although slowly, and recovers from heat damage at chill temperatures. It is the commonest cause of death (estimated to be about 100 annually in the UK in 1989 and 400 in the USA in 1990) from food contamination. (Of course, several other types of food have caused listeriosis, but one of the most notorious is cook–chill pâté – the wrapped supermarket type.) Readers, in the light of our knowledge of *L. monocytogenes*, might like to consider the following three simple proposed premises before the introduction of a new type of food.

1 The system should be able to restrain the growth or reappearance of food poisoning bacteria.
2 There should be a margin of safety to allow for inevitable defects in equipment and temperature gauges and human error.
3 There should be proper evaluation of the system before its introduction.

At present, cook–chill catering fails on all three counts.

Listeria monocytogenes is not the only psychrotroph* of concern. Others are listed as follows:

Organism	Minimum growth temperature	Notes on disease
Clostridium botulinum type E	+3.3 °C	Mainly found in fish, can contaminate other foods, causes botulism. Still rare in the UK, but some catering techniques (e.g. *sous vide*) may cause increase in disease
Yersinia enterocolitica	+1 °C	Causes acute or subacute illness with diarrhoea and abdominal pain, particularly in children, sometimes leading to surgery, being mistaken for appendicitis. Rare in the UK at present, more common in USA. Thought to be acquired from food such as pork products or milk
Enterotoxigenic *E. coli*	+4 °C	The cause of many cases of traveller's diarrhoea and enterocolitis in children. Common, but only rarely fatal. Transmitted through a wide variety of food and drink
Aeromonas hydrophila	−0.4 °C	Acute diarrhoea from eating many types of food contaminated with the bacterium
Listeria monocytogenes	−0.1 °C	Maternal–foetal infection, septic shock and meningitis, mainly in patients with underlying immune defects

Buying cook–chill food

The new UK 1990 Food Safety Act itself contains no times or temperatures for storage of packet food. But the new hygiene regulations and their numerous amendments issued from July 1990 *do*. For most cooked packeted food these regulations state that it should be stored below +8 °C, but there is no time limit at all. The retailer can therefore decide what he wants the sell-by date to be to suit his convenience. But he can be prosecuted under the general offence stated in this Act for selling food that is 'unfit for human consumption'. Even here there is an ingenious get-out clause in the same Act, that states 'It shall be a defence

*Ability to grow when cold.

for the person charged to prove that he took all reasonable precautions and exercised all due diligence to avoid the commission of the offence.' A good defence would be that he was complying with the hygiene regulations.

No wonder few major food retailers are prosecuted by local authorities. It is almost as if the supermarkets wrote the Food Safety Act to further their own bad practices. So in practice it is the small retailers who are prosecuted successfully because they cannot afford the costs of an effective legal defence. This further damages their image which in turn advances that of the supermarket.

The responsibility for food safety in the USA is the Food and Drugs Administration (FDA). This statutory authority has substantial powers to prosecute and confiscate whole batches of food, if any product were found to be contaminated. I am sure it would not hesitate to prosecute any company, however big or influential, if it was justified. Indeed, prison sentences for those responsible for selling contaminated food are well known and should be a major deterrent.

Two UK egg products

I wouldn't normally criticise particular products, but several colleagues have been so incensed by these egg packages, that I feel obliged to describe them.

The first is a single serving of a chilled cheese and ham omelette. The outside cardboard wrapping says it is 'ready to eat in five minutes'. There is no artificial colour, so presumably the eggs used were not laid by hens fed on the artificial orange dye, canthaxanthin, to make the yolks dark. Inside the wrapping is a sealed plastic container with two compartments. In one of these is fluid homogenised egg, and in the other a mixture of grated cheese and a few tiny strands of ham. The statement on the packet 'less than 10% meat' is evidently accurate. The packet then tells you how to make an omelette. It also says 'keep refrigerated', which is a clear indication of the introduction of unnecessary risk from processing and storing the egg – that is apparently not pasteurised. The cost of the ingredients is around 25 pence. The price to you is 99 pence. I wonder how much of the price is due to the cardboard and plastic, and how much is profit for the processor, distributor and retailer?

The second item, also making one portion, was described as scrambled egg and smoked salmon. The chilled cardboard packet tells you to preheat an oven to 180 °C and then heat the food for 18 minutes. Alternatively, instructions for using a microwave are given. The plastic dish inside the cardboard contains 7 oz. very sloppy scrambled egg, and ½ oz. smoked salmon. There is no claim for freedom from artificial colour. The cost of the ingredients would seem to be around 20 pence each for the eggs and smoked salmon, with a few more pennies for the butter, cream and seasoning. Say, a total of 45 pence. The price to you is £1.69, and you are also instructed to keep it refrigerated.

There is another way to make scrambled egg. I explained to my 7-year-old daughter, Gemma, what to do: 'Put the small saucepan on the cooker, add a dollop of oil and milk, a pinch of salt and a sprinkling of pepper. Turn the heat on about half-way, break two eggs into the saucepan, stir vigorously with a wooden spoon for a few seconds, then intermittently until the scrambled egg is as firm as you want'.

She managed it perfectly first time, and the dog had a delicious breakfast. The total time for the whole exercise was exactly 5 minutes, but of course we did have the diabolical hassle of washing up one small saucepan!

Convenience meals do have a role for the hard-pressed working person but these products don't save time, are wasteful on packaging and money, and must be potentially more hazardous than home-made meals. It is well known that once the yolk and white of an egg are mixed, some of the natural resistance to bacteria is lost.

Nutrition of cook–chill foods

The nutritional content of meals occasionally eaten may not be important. However, if a catering system provides the exclusive source of nourishment for an institution, then its nutrition is

important. That there is a need to supplement cook–chill food
with vitamins, etc., does suggest concern over some nutrients.
Because no study has been published assessing the nutrients of
the entire cook–chill system relative to other systems used to
feed a community, it is possible to point only to concerns rather
than to certainties. The first concern is that of the amount of
polyunsaturated fatty acids. Unfortunately these fatty acids fare
badly in cook–chill operations, with a tendency for decomposition under cold storage with not only loss of active compound,
but the formation of unpleasantly smelling derivatives. Thus foods
such as fish, and many vegetables widely thought to be desirable
for nutrition, are not at all easy to process by the cook–chill
method. It seems extraordinary that having identified the desirability of polyunsaturated fatty acids, catering systems are being
developed that reduce the content of these substances. Other
concerns include loss of folic acid (see Chapter 15).

Taste and smell

This must be the most difficult aspect to investigate, but common
sense must suggest that food produced by cook–chill methods
is inferior generally to conventional catering. The last cook–chill
meal the author was offered was a dish consisting of chicken legs
in a white sauce. The bones projected from the meat and were
grey-black, the meat was dry and tasteless and the sauce was
unpleasant, insipid and gelatinous. The vegetables were limp,
soggy and tasteless. Claims that cook–chill improves quality are
usually based on the fact that in hospitals reheating at ward level
can result in hot food. If cook–chill food is actually, from the
point of view of taste and smell, preferred to conventional food,
then there must be something seriously amiss with the latter.

Conclusion

Cook–chill catering is inherently more dangerous than cook–
freeze systems. The massive investment and commitment to this
system by the retail sector and by institutions has occurred largely

without proper consideration of listeria and other bacteria that flourish in the cold. Operations now in use must be controlled with meticulous care.

The advantage to the consumer is food available in ready-made form that can be heated within minutes from the chilled state. These may be the ultimate in convenience meals, and are not that expensive. But the environmental impact of transport, energy and packaging is very substantial.

Food irradiation

Irradiation of food has raised violent emotions, with the food processing and retailing industry usually enthusiastic, and the public hostile. I suspect the public has got it right, but for the wrong reasons! The advantages for the food processing industry are straightforward: it can enable contaminated food to be rescued and therefore used rather than dumped, and it can remove the spoilage bacteria from, say, chickens, so allowing them to last far longer on the supermarket shelves.

What happens when food is irradiated? Gamma rays damage bacteria and viruses, but not the agents that cause BSE and similar diseases. Electron beams are used in France and have an effect similar to gamma rays.

Radioactive cobalt-60 is usually used to generate gamma rays. This isotope spontaneously converts to nickel-60, with a half-life of 5.27 years – meaning that after this time the potency of the irradiation is half what it was, and after 10.54 years a quarter. If an accident occurred in the irradiation plant the radioactivity would stay around for years. The foods closest to the radiactive source receive more irradiation than those that are farthest away. The most uniform dose of irradiation occurs by treating a small item a long distance away from the source. This means that most of the irradiation is wasted, and it increases the cost and danger. This uneven potency remains one of the chief problems of irradiation and is particularly important because the differences in the amount needed to destroy microorganisms and that causing unpleasant tastes are small.

If bacteria are present in the food, they are killed because of damage to their DNA. The DNA molecule is a large, complex, double-stranded chemical with the sections joined by links between a sugar and a phosphate molecule. It is thought that these links are broken by the irradiation so that the DNA cannot function. Bacteria are pretty resistant to the effects of irradiation because of their ability to repair their damaged DNA. For example, the hole in one strand of DNA can be repaired by using the intact other strand as a type of template. Sometimes the repair is faulty with the new DNA being abnormal. New DNA means new properties. This is called a mutation, and one fear from irradiation is that it induces new, nasty, mutant bacteria. I wonder?

Amongst bacteria there is enormous variation in their vulnerability to irradiation. Those most easily destroyed are the food spoilage bacteria, which explains why irradiation can be used to prolong the shelf life of a chicken from, say, 1 to 2 weeks. One problem with these spoilage bacteria is that some can grow in the chilled conditions operating in supermarkets. However, surely such bacteria are an invaluable guide to the safety of food? If many spoilage bacteria are present they alert us to the presence of dangerous ones. Unfortunately, most dangerous bacteria do not have any smell.

Other bacteria that are easily removed by irradiation are salmonella and campylobacter, two of the notorious causes of food poisoning. But others, which may be dangerous, need much higher doses and may survive or even be encouraged by removing the competition. Others are completely resistant, including those that are able to survive hostile environments, such as drying, by converting themselves into tiny tough spheres. One of these is *Clostridium botulinum*, which when actively growing can produce a powerful poison that causes the disease botulism.

Unfortunately, the spores of *C. botulinum* survive doses of irradiation used in practice. Experiments have shown that irradiation can indeed encourage the growth of *C. botulinum*. *Listeria monocytogenes* is also fairly resistant to irradiation.

Another problem with irradiation is that the exposure of a food

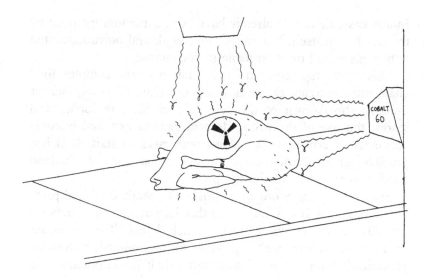

occurs briefly, over seconds or minutes, whilst other methods of preserving food, such as drying, salting, pickling, sugar, canning and deep-freezing all act throughout the storage life of the product. The problem is that after irradiation and subsequent storage, many things can go wrong.

Gamma rays might well produce dangerous chemicals from interaction with the packaging, and those chemicals be eaten subsequently. This is almost impossible to research adequately.

Poisons or toxins produced by certain bacteria after growing in the food are not removed by irradiation. These substances can then cause serious illness on their own account. So irradiation might destroy the bacteria that had already produced the toxins, but if those toxins are still present, the food is still dangerous. In practice, if a food consignment (for example, prawns) were found to be heavily contaminated with bacteria, irradiation could be used to mask the obvious evidence of contamination – that is, the presence of the bacteria – but would leave behind the toxins.

Exposing a food item to irradiation requires a journey to and from the irradiating plant, which must anyway increase the age of the product by the time of sale. If retailers then keep the item for longer than usual, the extra age may have an important deleterious effect on nutrients and vitamins. It won't be improved.

In any case, there will already have been some loss incurred by the irradiation itself. Vitamin C, folic acid, and polyunsaturated fatty acids would be most liable to decompose.

This extra step required by irradiation in the complex food chain must increase the risks from accidents, mistakes, etc, for example, breakdown of the refrigeration plant in lorries and stores, the use of the wrong dose of gamma rays and possible accidents or crimes at the plants with risks to staff. It is just possible that an accident could transfer radioactivity into the food itself. It is this that worries the public.

Irradiation is therefore an unsatisfactory method of food preservation, and it is not surprising that its potential use, benefits and advantages are still subject to much debate. Those sections of the food industry with a positive attitude towards irradiation presumably intend to use it to prolong shelf life of poultry or to clean up contaminated food. The benefit to the consumer seems difficult to identify.

As far as safety of eating irradiated food is concerned, then, there are absolutely no experiments that can be used to reassure anxious people. The only sensible approach must be to avoid these products altogether. There is no need to eat food treated with gamma rays.

At the time of writing in 1992, both the USA and UK governments have authorised the use of irradiation for certain foods. But many American states have declared opposition, as have the great majority of consumers. In practice, I have not been aware of any substantial use of this process in either country. However, persistent rumours abound regarding its use elsewhere in the world and the import of unlabelled irradiated food, such as strawberries.

Vacuum packs

Many processed foods – and also some unprocessed ones, such as fish and meat – are sold in containers with either a vacuum (i.e. an absence of air altogether) or with special gases blown in.

The main reason for these is to stop discoloration of the foods and also to keep them longer before being sold. There is nothing unsafe with these products as long as they are stored really cold and not for too long. The important point to realise is that a vacuum itself is not a preservative. The absence of oxygen or indeed general alterations in gases tends to stop some bacteria from growing, but encourages others. Some of those encouraged might be dangerous.

So treat food purchased in vacuum packs or with modified atmospheres (if you can recognise them!) as any other perishable food. If you don't fancy the colour, feel or smell, do not eat it. Follow any storage instructions to the letter and make sure your refrigerator is really cold (0 to +3 °C) if you store these. As with so many products, we don't need vacuum packs, and they must add confusion and potential danger.

Canned food

Canning should now be a satisfactory method for storing many food items. True, there may be loss of Vitamin C and the quality is not the same as the real thing. But canning has been well researched; we have learnt from our mistakes. It usually involves heating to around 130 °C, the temperature required to kill all bacteria. Ham is spoiled at this temperature, so it is heated to a lower temperature and a preservative, usually sodium nitrite, has to be added.

So whilst canned goods are just about as safe as any other, when accidents rarely occur, they are pretty spectacular. These result from the failure of proper heating so the bacteria that cause botulism can multiply, even in the sealed can. This used to be a fairly common problem early this century with the fashion in the USA for preserving vegetables in the home by canning. But in the UK, disasters from canning are very uncommon. Over 60 years ago, a party fishing on Loch Maree in Scotland succumbed to botulism from duck pâté. Then, in 1978, four people developed botulism from eating inadequately heated canned

Alaskan salmon. More recently, in 1989, about 27 people were poisoned by yogurt from the addition of defectively canned hazelnut purée. Amazingly, the cans were obviously bulging from the build-up of internal gas, yet the yogurt makers still used them! These are the exceptions to the rule, and only four people are known to have died from botulism caused by canned foods in the UK in the last half-century.

A few simple checks should be made when buying cans. First, the ends should not be bulging outwards; secondly, there should be no rust holes or severe denting; and thirdly, the joins in the can should be intact on removal of the label. If there is a hiss when the can is first pierced, this is usually due to the vacuum inside, rather than the escape of a build-up of gas.

As long as the can is intact, it should be safe almost indefinitely, although the quality may deteriorate after some months. It is quite in order to keep the can at room temperature. Never put a can in the deep-freeze, as the contents could expand and burst the seam. Perhaps the only common danger comes from the opened can, that should be kept in the refrigerator for not more than 2–3 days. I believe there is no general need to transfer the contents after opening to another receptacle.

So what canned foods can be positively recommended? Those of us with an unfulfilled yearning for beef can safely eat canned Argentinian corned beef! I like anchovies in olive oil, although they do seem rather salty. But I simply cannot comprehend the popularity of mushy peas. Perhaps living in Yorkshire, UK for 9 years is not long enough for me to appreciate their subtle qualities! Millions of cats and dogs seem to thrive on canned food.

Low-fat margarine and spreads

Our diet-conscious society has an insatiable demand for low-fat margarine. What is put in margarine to make it low in fat? The answer is water, mainly. Indeed, some spreads contain more water than fat and it could be argued that they should be described as 'high fat water'! They are certainly a very expensive way of buying water!

The basic ingredient tends to be vegetable oil, and the various constituents are aimed at producing a stable, thick water/fat emulsion for spreading (but not for baking). The addition of water introduces the threat of bacterial growth, even under refrigeration, so most contain preservatives – sorbic acid or potassium sorbate (these do seem safe).

Recently I was introduced to a new brand of margarine, with a claim that it was, to the consumer, indistinguishable from butter. Certainly it looked like butter, and had a mild, even pleasant, smell and taste. Yes, it could be mistaken for butter. It was also cheaper than available brands of butter.

But the crucial question is: is this product better for you than butter? The first point to remember is that most nutritionists tell us that many of us are on average eating too much fat, whether it is butter or margarine. Indeed about 30% of our food by weight is still fat. Then, there is evidence that fat high in saturates obtained from animals is less desirable than fat high in polyunsaturates obtained from fish or vegetable sources; we will look at this again in Chapter 15. It is still not certain whether animal fat is particularly undesirable because of the high amount of saturates and cholesterol or the shortage in polyunsaturates. To make the matter more complicated, most fats contain a proportion of mono-unsaturates, best viewed as half-way between saturated and polyunsaturated.

So, what actually was in this margarine? The manufacturers would not tell me the source of the vegetable oil, but they admitted that up to 25% of the fat had been hydrogenated, and the final product contains around 34% natural polyunsaturates, 15% mono-unsaturates, 7% of the unnatural *trans* fatty acids, and 17% saturated fats. The remainder was mainly water. So the key question is, does the 7% *trans* fatty acid produce hazards that more than outweigh the benefits of some of the other products? I think they might do, but research would be almost impossible to prove or disprove this.

We seem now to have a dilemma. If we do really want low-fat spreads, we will have to use more saturated fat or hydrogenated oils in order to keep the texture adequately solid. The alternatives are a vast array of chemicals. Is this not complete nonsense? Do not low-fat spreads waste a huge number of containers? Do you not approve of the move to more concentrated detergents? Why the move to purchasing more and more containers of expensive water? Surely the answer is to eat less high-fat fat, rather than more low-fat fat?

The food industry is correct when it claims it provides the consumer with what he or she wants. So it is up to you.

Salt, sugar and vinegar

Salt and sugar preserve food through their osmotic action. What happens is this. Suppose a bacterium alights on the top of, say, jam; because there is so much sugar in the jam, the bacterium's own water gets sucked out into the jam so it cannot grow. The key for sugar and salt in acting as food preservatives is the amount present. Sugar substitutes won't do and if public concerns about eating too much sugar and salt continue, then the use of these preservatives will decline further.

It is simply not possible to reduce their concentrations: if you try, bacteria will start growing once more. One reason for the use of so many chemical preservatives has been the fashion for low-salt and low-sugar everything. Incidentally, fungi can grow on sweeter foods than can bacteria, so home-made jam can become contaminated with fungi.

Vinegar kills bacteria by acidifying them to death, and the only problem is the taste. But vinegar is completely safe for us. A great deal of nonsense has been written about the danger of vinegar, on account of its acidity, but our own stomach fluid tends to be more acidic than preserving vinegar!

It is a pity that the fashion for reducing the amounts of preservatives has spread like a contagion to vinegar products. If, for example, the amount of vinegar is reduced in mayonnaise, then in order to stop bacteria growing after the jar has been opened (it can be made safe through heat initially), it must be stored in the refrigerator. There is nothing wrong with old-fashioned pickling!

Meat

Pork pies

Pork pies are particularly popular in the UK, but less so in the USA, and are probably the most effective means of providing a very high intake of saturated animal fat. The pastry is about 50% animal lard, and the minced pork gunge is often mainly fat. The gel, however, contains protein, usually made from beef bones.

I can see some attraction in the freshly baked pork pie. The pastry has a seductive smell and the inside is definitely savoury. Quite useful for a picnic. Acceptable as a luxury – just. But what about the dreadful product after being stored cold in a supermarket for a week? Has it any taste at all? Do you eat it only because it reminds you of how the real thing used to taste?

The supermarket pork pie represents everything that is rotten with our food.

Other meats

Most ready-cooked sliced meat has been cooked in large joints enclosed in polythene wrapping and steamed for several hours in chambers. This method of cooking is aimed at keeping the water content of the meat so that it can last for several days after slicing.

After steaming, the outside of some joints, and in particular of whole chickens and turkeys, is smeared with caramel, often with other sugars and salt to give an impression of roasting. These products should not be described as roast, but just cooked. Cooked meats are notorious for their ability to support growth of bacteria, and unless very rigorous precautions are taken, they can be heavily contaminated. Several outbreaks of salmonella food poisoning, and also cases of listeriosis have been traced to these products.

The problems result from the uneven penetration of steam between tightly packed joints, so even if an inserted thermometer indicates, say, a final temperature of 80 °C in the centre of a joint, it cannot be assumed that this has been the case for all the rest. But it is interesting to see in the USA cooked chickens for sale with meat thermometers inserted in their flesh and reading temperatures of 150 °F (74 °C).

But a greater problem than this is the transfer – usually by human fingers – of bacteria from raw food to cooked. Take a close look at retailers of cooked and raw meat, and you will see what I mean.

Pâté is generally made in two ways. In the first, the liver and other animal components (including the unspeakable!), after conversion to a purée are subjected to the canning process, and should therefore be microbiologically safe.

The other approach is a variant of the cook–chill, where the pâté is heated to around 70–72 °C, chilled and wrapped, and stored cold. Listeria has been found in some of these products and is due either to its ability to survive the initial heating or to its accidental contamination later. In either instance, the prolonged cold storage can allow the listeria to multiply. Such centrally produced food, often from another country, and its tortuous distribution, seem fraught with risks. Of course, locally made and genuine fresh pâté is a different matter. There are still a few people who make it themselves. In December 1992, the Department of Health reissued its advice that pregnant women should not eat pâté, or soft crusted cheeses or cook–chill food unless it was thoroughly reheated.

Salami is made through fermentation, and if care is taken in the initial seeding of the vats to ensure freedom from contamination, there should be few problems. The product of the fermentation is acid rather than alcohol and the acidity is a reasonable preservative. But the products are very fatty.

Yogurt

We are now experiencing a major epidemic of yogurt! Despite a drop in popularity in 1989 in the UK as a result of reports of botulism after eating the hazelnut type, the yogurt epidemic marches on and is now second only to the crisps outbreak. Brands in my local supermarket included Natural low fat, Greek strained, German, Thick and Creamy, Forest Fruit, Live Natural, Light Mousse, Virtually Fat Free, Extra fruit, Natural set, French, French style, Mandarin, Raisin and Rum, Hazelnut, Fiendish Feet, Lightly Whipped, Dessert, etc., etc.

Yogurt is made from milk. First the milk is heated, usually to higher temperatures and far longer than with pasteurisation. Then bacteria are added that cause the thickening and acidity that act as a preservative. The bacteria added are called *Lactobacillus bulgaricus* and *Streptococcus thermophilus*. They change the lactose in the milk to lactic acid which causes the previously dissolved proteins to precipitate, accounting for the thickness of the yogurt. The idea that yogurt is natural and contains no preservative is clearly misguided. The nutritional content of yogurt might well be inferior to that of milk, because of the additional heating, bacterial activity and storage. As long as the yogurt is prepared under superclean conditions, and is transported and stored cold, it should be safe microbiologically. However, yogurt need not be stored at temperatures low enough to keep, say, cooked chicken safe, so the risks with yogurt must be lower. Moreover it does contain a preservative, acid.

The problems with the hazelnut yogurt came from the hazelnuts, and let us hope that manufacturers have learnt the lesson that every component of a processed food item must be carefully checked.

So, enjoy your yogurt. Ignore advertising slogans – Fresh, Natural and Free of Preservatives: these have no meaning. It is up to you, if you want to buy what is effectively milk at five times the prices of the unprocessed product.

Cheese making

Most cheese is made from pasteurised milk, but it certainly need not be, and many connoisseurs rightly argue that the better flavours are derived from unpasteurised milk. The important factor is that in the early stages of cheese making, acid is produced that is likely to destroy any extraneous bacteria present in unpasteurised milk.

Like yogurt, bacteria are added to milk to start the process off. Rennet is added to clot the proteins, and the solids separated from the liquid whey. Now we have cottage cheese. This is further treated, stored, and sometimes impregnated with wires, and then ripened.

Let's take soft crusted cheeses such as Brie and Camembert and see why listeria used to be found in as many as 10%. I say 'used to be', because some of the makers do seem at long last to be cleaning up their act. Raw, unpasteurised milk may rarely contain small numbers of listeria, but this is not the main problem. The cheese making process is started by adding millions of harmless bacteria to milk, and checks are made that no unwanted, possibly dangerous, bacteria are added. The bacteria produce acid, the milk proteins solidify, and the cheese gradually matures in a cool atmosphere. Initially the cheese becomes so acidic that listeria and other dangerous bacteria cannot grow. During ripening a crust or skin forms on the outside of each cheese and the acidity lessens. If the ripening cheese is then contaminated by environmental listeria they may now start growing.

Contamination can also occur at the point of sale. Ideally soft cheeses should be loosely wrapped in an impermeable material and placed in a rigid box to prevent bacteria entering. It is always preferable to buy whole soft cheeses rather than portions because cutting into cheese can let in bacteria.

For ripening the cheese should be kept cool – say, at +10 to +12 °C – or in the refrigerator if a cool room is not available, and eaten as soon as possible after being cut – say, within 2 days.

Preventing listeria getting into cheese needs meticulous hygiene at all stages in production, transport and sale. Not long ago, we were puzzled why certain cheeses from a retail outlet were often contaminated with listeria. A surprise visit to the shop revealed the cause: the cheeses were standing naked on dirty straw!

I am afraid that we still must be cautious over these cheeses. Healthy young adults may find the small risk acceptable, but pregnant women, the more elderly and those with any illness should stick to hard cheeses.

PART III
Composition of food

8

Meat substitutes

Introduction

This chapter is for vegetarians, either those already committed, or those aspiring or interested. It is not the intention here to persuade all readers to become strictly vegetarian. Rather, it is hoped to show that those people who do not wish to eat meat or fish can still eat an enjoyable and nourishing diet. Vegetarianism has moved away from the eccentric fringe of society to being a very reasonable and respected way of life.

Our preoccupation with meat is very much a recent phenomenon. The idea that a meal *is* meat and two vegetables had seemingly become part of 20th century civilised man. Meat is also, in absolute terms, cheap as a result of intensive rearing and 'greater efficiency' (Chapter 3).

The purveyors of meat have also succeeded in excellent propaganda about the beneficial effects of eating meat. In fact, I know of no research that establishes any beneficial effect of eating meat rather than other food on our long-term health, except earlier this century, when Vitamin B_{12} from raw liver was found to be beneficial to people suffering pernicious anaemia. Even here, raw liver has been replaced by injections of purified B_{12}. Indeed in 1990, the UK government actually warned pregnant women against eating liver because it might cause defects in babies on account of its excessive content of Vitamin A.

There has been an extraordinary belief that a meal should consist of a central feature of meat or fish and that vegetables are of trivial importance, mainly for adornment. This is certainly the case for the increasing amount – and increasing amount that is not eaten – of salad garnishes.

It is in practice perfectly feasible to avoid meat altogether, and as long as the items of a vegetarian diet are chosen carefully, it could actually be preferable to meat-based food. We will return to this in detail.

It is almost unnecessary to comment that vegetarians are concerned about the practice of rearing animals for food, the cruelty and the slaughter. I believe these emotions are common to most people, the exceptions being mainly members of the male sex, some of whom hunt foxes and shoot birds for fun.

That only a smallish percentage of the population is vegetarian is because most people are able to dissociate their concerns for the live animal from the item on their plate or in their fingers. This is particularly easy to achieve with processed food that has no physical resemblance to its original state, and probably accounts for some of the popularity of processed meat. How many know or want to know the source of the meat in pâté or salami? Let's be even more sceptical: how many people know that cheeseburgers contain beef?

Meat products, particularly those processed and precooked, cause food poisoning very much more often than vegetable food.

Then there is the problem of the efficiency of conversion of food energy and other uses in man. Each time food is eaten by a mammal or bird, the process whereby the nutrients are converted into useful energy or growth is inefficient. At best only 30% of the potential food value is actually used. At worst, as with cattle grazing, it is only 7%. So if wheat is fed to chickens, whose meat in turn is eaten by us, at least 70% of its food value is wasted, compared with what would occur were we to eat the wheat directly as bread.

It is true that some areas of the world such as uplands can only be used to feed, say sheep, through grazing. This is not in question. What is disputed is the desirability of growing cereal crops on lowland specifically for feeding food animals.

Meat

Meat is the muscle of the mammal or bird. It is made up of long fibres assembled in small bundles held together with sinewy material called connective tissue. The small bundles are aggregated into larger pieces that have specific mechanical functions and to achieve these are inserted into bones through tendons. Much of the meat we eat is derived from muscles responsible for the animal's ability to walk.

A comment on the colour of meat is appropriate here. The colour of fresh meat is due to the presence of myoglobin, the protein involved with muscle contraction, which is purplish-red. Atmospheric oxygen tends to make the dark colour become a brighter red as oxymyoglobin is formed, but on storage in the absence of much air the colour reverts to muddy purple. Some of the gases put into packs of meat are aimed at keeping it an appealing cherry red colour. Almost exactly at 60 °C the red colour of the myoglobin is converted to a brown substance called hemichrome. So if this colour change occurs during cooking, then most bacteria can also be assumed to have been killed.

If the chemical sodium nitrite is added to meat, a new chemical

Where's the cooked **Myoglobin** main course to go with this **vegetable** side dish Mam?

is formed that is pink; this accounts for the pink colour of ham and bacon. Pork from the same animal not so treated is pale brown.

The average protein content of red meat is 15%, that of fat (mainly saturated) 25–30% and water 55%. Some vitamins and minerals are present. There is little Vitamin A or D, or folic acid (one of the B vitamins). There is also little Vitamin C and virtually no Vitamin B_{12}. A particular deficiency in meat is fibre: there is none. So you cannot live on steak alone: its nutritional value has been greatly exaggerated by its producers.

Below is shown the typical nutrient content of wholemeal bread compared to typical red meat:

	% by weight	
	Wholemeal bread	Red meat
Protein	10	15
Carbohydrate	40	1
Fat (total)	2	25
Fat (saturated)	0.5	12
Fibre	8	0
Water	35	55
Presence (+)		
Vitamin C	++	±
Vitamin E	++	±
Vitamin B complex	++	±
Calcium	++	±

The differences are striking in some ways, with the bread higher in complex carbohydrate (starch) and the meat in water and fat, mainly saturated fat. Perhaps the most surprising comparison is that of the protein content, with the bread containing as much as two thirds that of the meat. However, the type of protein in wheat has been considered less desirable than that in meat because it lacks of one of the essential amino acids, lysine. But this is easily obtained from eggs, cheese, milk or beans. The vitamins in the bread are partly 'natural' and partly added.

The obvious conclusion from this comparison is that the nutrients of wholemeal bread seem preferable to that of meat (see also Chapter 15). The concept that meat is a required or ideal 'all round' food is totally misplaced. It is true that meat is an excellent source of iron and protein; but so are many other foods. Meat is relatively high in saturated fat and deficient in fibre.

The point is made, I hope, that there is no requirement for anyone to eat meat.

Fish

The argument against eating fish is not as compelling as that against meat.

Most fish contain around 15–20% protein, and between 2 and 20% fat, but unlike meat fat, this is essentially unsaturated (but see Chapter 4); they contain virtually no carbohydrate, nor fibre. However, fish are generally a good source of vitamins A, the B complex, and D, and also calcium.

There are disputes as to what extent fish can suffer pain: certainly their nervous systems are much more primitive than those of mammals. One can therefore understand the views of people who, for example, will not eat meat but do eat fish, and of those who do not eat intensely reared meat or fish, such as farmed salmon.

Meat substitutes

Immediately, we can identify a problem with the concept of meat substitutes. The idea that meat is a necessary nutrient and that vegetarians because of their emotional concerns may suffer from nutritional deficiencies is not true, at least only very rarely.

The concept of developing a product that looks like meat, tastes like meat and is similar in nutritional content to meat is quite difficult to achieve, and may well not be what many vegetarians actually want. This is not to decry the product altogether.

Consider a group of diners drawn to a burger outlet like a giant magnet attracting iron filings. What does the vegetarian amongst them eat? Yes, a vegeburger or nothing! Many other vegetarian meals are so unattractive after storage chilled or frozen and reheating that the vegetarian burgers become almost desirable – relatively speaking.

But were baked beans on toasted wholemeal bread available, then surely the vegetarian would prefer these?

Tofu

One of the substances used as a base for meat substitutes is tofu, which is extracted from soyabeans. It possesses many of the desirable features of pulses: high in protein, low in fat and rich in minerals and some vitamins, such as the B complex and E. It is said to have been used for 2000 years by the Chinese and the various derivatives of it are available in many retailers.

For example, to make Tofu burgers, I am indebted to the *Vegetarian Magazine*.

TOFU BURGERS

Ingredients

1 block/285 g/10½ oz. firm tofu
50 g/2 oz. bulgar or
100 g/4 oz. cooked millet or rice
2 fl. oz./50 ml hot water
1 small onion, very finely chopped
50 g/2 oz. mushrooms, very finely chopped
50 g/2 oz. carrot, grated
1 vegetable stock cube, crumbled

2 tbsp. soya sauce
1 tsp. dried mixed herbs or
2 tsp. any fresh herb, coriander, dill, etc.
2 tbsp. wholewheat flour plus a little extra for coating

Method

1 Break up the tofu with your fingers into very small pieces.
2 Mix all the ingredients together in a bowl and squeeze well to shape into burgers. (As the mixture is very sticky it will help if you dip your hands and utensils in cold water.)
3 Refrigerate for an hour or two.
4 For a crisp coating, toss the burgers in a little wholewheat flour before frying.
5 Deep fry the burgers in oil, a few at a time.
6 Drain well and serve hot or cold.

It is most important to fry the burgers in very hot oil to avoid disintegration.

Other types of tofu include silken, extra firm, and smoked. You can use it as minced meat substitute, in sweets, in 'rissoles' or in various pasta dishes.

How does the protein of tofu compare with that of meat and fish? Adults need about 10–20 g protein each day under most conditions, and proteins in our diet contain the small parts, or amino acids, that are incorporated into the body's protein. But they do more than this. They provide usable nitrogen which can, for example, be needed for making DNA. The body can make some of its own amino acids from others in the diet, as long as there is enough in total, so proteins in our foods really have two functions: one is the supply of essential amino acids that we cannot make, and the other is a general provision of nitrogen that we can use.

This means that proteins vary in their quality and potency, or if you like, in providing us with these essential amino acids.

Some nutritionists place proteins on a scale, with those foods that contain amino acids that match our specific needs given a high score, and those that contain few essential amino acids a low score. Eggs are top with a score of 97, milk scores 83, meat and fish 75 and soya 65. The latter score is relatively good, and certainly a mixture of tofu with eggs or milk should be equivalent to meat or fish.

Apart from tofu, soyabean protein is used in a variety of other types of imitation meat products. So too are extracts from fungi and bacteria. I do have a criticism of many of these products as they are available. They are often available in outlets that do not possess the proper facilities to store them. I have seen vegetarian sausages and pâté stored at room temperatures when they should have been refrigerated.

Some vegetarian meals served in restaurants are limp, tasteless and often very ancient. It is as if the restaurant is making a gesture to the odd crank and feels obliged to store a variety of vegetarian meals to await the order, that may not come for days or weeks. The restaurateur does not often put his 'heart' into this type of food unless he specialises in it.

Of course some vegetarian food at retail outlets and restaurants is excellent, but the central mistake does seem to me to approach vegetarian eating by copying the shape, colour and taste of meat-based food. Why not enjoy vegetable food for its own sake? We come back to the 'meat and two veg' syndrome.

Many vegetarian products are made in the image of meat to provide a central focus for the platter. The beanburger replaces the beefburger, the vegetarian rissoles with textured soya protein look and taste like pork. The vegetarian pâté looks and tastes like chicken liver pâté. But wait a minute. Do vegetarians want to be reminded of meat when they eat? I wonder. Surely there is a great misunderstanding that the nutritional content of meat has somehow to be re-created or imitated. I will say it again. We do not need to eat meat.

The problem for vegetarians is often to produce a main feature for a meal around which accompaniments help promote the importance of the feature. How, for example, do you replace a giant slab of steak garnished with a few chips, onion rings and salad? Of course there are many easy ways of establishing a dominant structure to a vegetarian meal. Let me suggest a few.

The obvious example is a vegetarian pizza! Pizzas do naturally lend themselves to toppings of cheese, tomato, onions and/or peppers and chilli. Slices of meat never appear comfortable on a pizza base.

Pastry, often in flan form or as a type of turnover, can provide a positive focus. Alternatively, a vegetable dish itself can be featured. Cauliflower cheese with decorations is one suggestion. Various types of toppings based on egg products or beans placed on toast can easily be decorated to look really attractive.

Rice, too, such as a mixture of four parts long grain to one of wild rice with gratings of peppers can be so arranged as to provide a central feature; for example, this can be placed around the periphery of a plate with central 'vegetarian' casserole.

Many people's view of vegetarian food seems confined to the nut cutlet or beanburger, neither of which excels as a feature. Why chop nuts up into pieces? Why not enjoy an egg and leek flan with top decorated with slices of walnut and tomato halves?

I hope I have made the point: so much of vegetarian cooking is really dreadful because it attempts to simulate the colour, shape and nutrients of meat. It is based on the standard prototype 20th century meal of 'meat and two veg'. With a little thought and trouble, vegetarian food can be made both appealing and nutritious; and, of course, it generally less expensive than meat (see also Chapter 17).

9
Food poisoning

Introduction

What is food poisoning? There is no one simple definition or cause. It usually refers to problems from unwanted bacteria or viruses in the food that make you ill after eating it. But sometimes chemicals such as histamine in some tropical fish or toxins in shellfish can produce illness. Or it may not be due to the bacteria themselves, but to substances they have already released into the food.

There are two essential features of food poisoning. The first is that it produces illness in the short term; perhaps listeriosis and hepatitis 6 weeks after eating contaminated food is the longest interval (this is known as the incubation period). The second point is that we can and should prevent it.

The causes of food poisoning are varied, and it often results from a combination of accidents or mistakes. For example, something may be amiss that lets dangerous bacteria into the food in the first place, and something else makes the numbers of bacteria rise sufficiently to make you ill after eating. So in apportioning blame, all the people involved in the food chain are responsible, maybe even you sometimes! It is regrettable to have to make this comment, but food poisoning would seem to be a bigger problem in the UK than in the USA.

We can begin with the contamination of animal feed that in turn contaminates the animal. The infection may then be spread around in the slaughterhouse or in preparation of the meat product for sale. Processing, particularly if involving mincing, introduces new infections or it may spread around the bugs already present. Transport, storage and retail sale can all produce problems, notably when refrigeration fails. Mistakes in restaurants, take-aways, and finally you, in the home, can also be responsible. The problems here can partly be due to human errors or to equipment faults, in particular from using the equipment in the 'wrong way'. Microwave ovens are an obvious example here. By their very nature of operation, microwaves heat briefly and unevenly and must pose risks.

In pointing the finger of guilt, it therefore goes to everyone. However, the initial contamination from unsafe farming methods would seem to be the biggest single factor.

Has food poisoning really been increasing, or is it all media hype? The first uncertainty is that the ability to detect and monitor the incidence of food poisoning varies from country to country. Unfortunately those countries that are believed to experience the worst problems, such as parts of Southern Europe, have the least effective procedures for assessing the scale of the problem.

There are two general approaches in attempting to quantify the problem. One is passive, in that it depends on the expectation that cases will be diagnosed and reported by the initiative of the staff involved. The other is active, when enquiries are made into all the possible cases, usually with intensive study (called surveillance) of a small but representative fraction of the whole.

Some countries are more able to detect food poisoning than others, and this is very much linked to the sophistication of their medicine practised. In poor countries, patients are unlikely to agree to pay for expensive tests that may not benefit themselves directly, although such tests may be needed to find the culprit food. Certainly there are differences in the ability of medical centres to identify bacteria, toxins or viruses.

However, there have only been small changes over the years in the way that cases are identified and reported, and whilst it can still be argued as to how many cases of food poisoning are being missed, there is little reason to think that the proportion has changed much. There is one exception: that due to the bacterium campylobacter, the isolation of which was beyond the ability of most laboratories before 1980. Some of the apparent increase in campylobacter infections is due to our improved ability to identify them.

No doubt media interest in food poisoning can appear temporarily to exaggerate its incidence, mainly as a result of making patients and doctors more aware of its possibility. In the UK most of the media attention occurred in late 1988 and 1989, and there may have been some distortion of the figures here at that time, but not since.

But, even allowing for these uncertainties, the conclusion has to be very pessimistic, with the total increase in real cases likely to be four-fold in the UK between 1982 and 1992. The increase in the USA has been about 2.5-fold over the same period. Other countries may not be able to document these trends so accurately, but there is every reason to believe that the upward trend is almost universal in the developed world. That other countries are as bad as – or worse than – yours in no way condones the situation. Politicians and the food industry please note.

The main 'events' of the last 10 years are as follows.

1982   The first proof of the link between chilled food and listeriosis. It occurred in eastern Canada.

1984   The two biggest salmonella outbreaks in the UK. One, in Stanley Royd, Wakefield, involving 455 patients and staff, was given maximum publicity through an official inquiry, and was followed by attempts to privatise NHS catering. The other involved 766 American travellers returning from Heathrow. This poisoning was associated with a cook–chill operation and received virtually no publicity, presumably because it might have been an embarrassment to the privatisation of British Airways that year.

1985   The beginning of the salmonella in eggs problem in many countries of the world.
        Epidemic of listeriosis from Mexican-style soft cheese in California, affecting 150 people with 46 deaths.

1986 The beginning of the BSE endemic in the UK.
1987 The food poisoning figures were now rising by about 30% per annum in both UK and North America.
1988 Nearly 100 deaths from listeriosis in the UK and 400 in the USA.
1989 The biggest botulism outbreak in the UK since records began. It occurred in NW England and resulted from contaminated canned hazelnuts being added to yogurt.
1990 Increasing concern by the public about cruelty to birds and mammals farmed for food. Salmon farming shown to be environmentally damaging and produces sub-quality fish, even though of a bright orange hue.
　　　Pregnant women in the UK advised that liver was dangerous.
1991 British beef continues to be boycotted by many countries.
1992 Epidemic of listeriosis in France, affecting hundreds of people and of uncertain source.

Shigella (dysentery)

'We will soon have a society not knowing how to use knives and forks.' I am not sure who said this, but it is depressingly true. Does it matter that we now eat with our fingers? Doesn't it save us all the hassle of washing up? Those polystyrene boxes are so convenient, and surely it is not *our* responsibility to dispose of them?

Will cutlery soon be the domain of antique shops? Perhaps we will start collections of silver, steel, plate, alloy, bone, wood, plastic and even cardboard types to commemorate the bad old days of having to sit down to eat, and having the nightmarish difficulty of transferring food from the surface of slippery plates with cutlery into our mouths. There is, in reality, a very good reason for using cutlery when eating, and the use of knives and forks rather than fingers is one of the main distinctions between a civilised and an uncivilised society.

You may have heard of the problem of dysentery among school children. The bacterium that causes it is called *Shigella sonnei* and the disease it causes is distinctly unpleasant. It is on the increase in many countries and the problem is aggravated during summer weather. The natural 'home' of shigella is people's intestines and not much elsewhere. So, for it to spread between people it has to leave one person's intestines and get into another's.

How can it do this? It gets out of the first person's intestines at the bottom end and because it passes through toilet paper it can get onto the fingers, and can persist invisibly for some time unless the hands are washed very thoroughly.

How does it pass from one person's fingers into another's intestine? Consider this. A group of children visit a fast food 'restaurant' after school. They can't afford a full bag of chips (fries) each, so they share. Fingers containing the invisible shigella go into the bag and contaminate the other fries. The next child picks out a contaminated fry and eats it. The greasy fry is chewed with the bacteria distributed into the centre of the lump of potato that is swallowed.

The fat protects the bacteria from destruction by acid in the stomach, and they then enter the lower bowel after 12–18 hours and increase there in numbers. Here they release substances that irritate and damage the bowel and so produce diarrhoea and severe colic.

This explains how the shigella bacteria can get from one person's intestines into another, why the disease is on the increase and why civilised societies use knives and forks. Incidentally, we now refer to this infection spread as through the four 'Fs' – faeces, fingers, food, face! Shigella is a major problem in the USA and the UK.

What food poisoning does to you

Food poisoning is definitely unpleasant. The most violent symptoms are due to eating pre-formed toxins.

Some people are more prone than others to suffer from food poisoning: the elderly, pregnant women, newborn babies, those with particular diseases, those on certain drugs and those whose immune systems are below par. Some healthy people can suffer severe symptoms, even from infections due to listeria or salmonella.

Salmonella food poisoning

This has been most studied and to many people salmonella *is* food poisoning. It results from the ability of the bacteria, after being eaten, to survive the stomach acid. Subsequently the bacteria start multiplying in the intestine. In most instances the illness begins 12–24 hours after eating contaminated food and rarely after 48 hours.

The symptoms result from the effect of the bacterium's poison (known as an enterotoxin) on the lining of the intestine so that water and salts collect in the middle of the gut. Colicky pain all over the abdomen, with urgent and explosive diarrhoea, are usually the first symptoms. The temperature may be raised to 39 °C (101 °F), but not always. Vomiting can occur. After a few hours, loss of water and salts in the diarrhoea, and a feeling of not wanting to drink, cause signs and symptoms of dehydration: sunken eyes, dry mouth, loss of skin firmness, small amounts of concentrated urine and a rising pulse rate.

Elderly people may be particularly poor at compensating for loss of water and salts from the blood. If so, the volume of the blood shrinks, the pulse rate goes on rising and the blood pressure drops. The entire circulation can fail and complications from low blood pressure can occur, such as strokes and heart attacks.

Fortunately, most people begin to recover within a few days, with the diarrhoea becoming less. However, full return to health can take a month or more. During this period, the patient may feel exhausted, or mentally and physically lethargic. Whilst recovery is usual, in about 1–2% of patients other complications occur, particularly in those who are ill or elderly, and result from the bacteria escaping from the inside of the intestine to the bloodstream. When the salmonella multiply in the blood, symptoms of general blood poisoning or septicaemia occur: high fever, e.g. 40–41 °C (103 °F), violent shivering, rapid pulse rate, low blood pressure, and wheezing. Admission to hospital and a correct treatment should lead to recovery.

The source of salmonella

Salmonella is found mainly in poultry, some mammals and even reptiles. In poultry, salmonella is found during the bird's life in a fairly small proportion of broiler chickens where it may or may not cause disease. It is the automated slaughtering and preparation that spreads the bacteria widely. At the point of sale, the majority of broiler carcasses will contain salmonella – perhaps anything from a thousand to a million bacteria on every chicken. Contact with other food will spread these bacteria further.

There are about 2000 different types of salmonella. These have one or more natural hosts in which they can live in the intestines without necessarily causing disease. Some salmonella are able to infect just one or two hosts, whereas others can involve a wide variety. For example, *Salmonella enteritidis* infects poultry and man, but *Salmonella typhimurium* is found in a wide range of birds and mammals.

With chickens, the advice of thorough cooking and avoidance of stuffing of the cavity surely need not be restated?

Eggs

There are two ways in which bacteria can get into the inside of an egg. Until recently, the best known method was through shell contamination and damage, usually after laying. The most obvious way that these bacteria can get into the egg is through droppings being forced into a crack in the shell. Shell contamination has been known for years and there is no reason to think that it has been on the increase. The other method is the direct spread of salmonella from the chickens' ovaries to the eggs as they are being formed. This is called 'transovarian'.

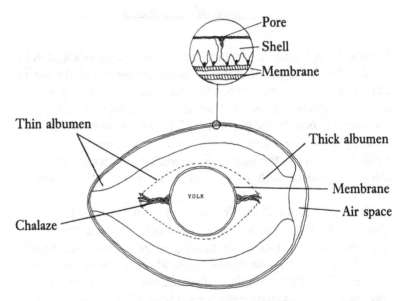

Inside structure of a hen's egg

Transovarian contamination of eggs

'Transovarian' means that the salmonella passes directly from the ovaries – the egg producing organs – into the inside of the eggs. The evidence in favour of this has come from research in the

UK and the USA. First, salmonella has been found within the intact eggs. Next, salmonella bacteria have been found in the ovaries of the laying hen; despite this, many of these birds had been laying quite normally. Then experiments have shown that feeding hens with *Salmonella enteritidis* can result in the bacteria appearing in the ovaries. In the USA, the strains involved are slightly different from those in Europe.

It is not possible to assess exactly how many eggs are contaminated with salmonella. In the UK it is of the order of 1 in 5000 to 1 in 10,000. This figure was estimated by the author in 1988; it still seems right. The degree of contamination is not uniform. Recent surveys from Devon, UK have found that about 1 egg in 200 is contaminated. Any one person might consider that the average figure of 1 in 5000 is unimportant, as it will take the individual consumer 15 years to eat one contaminated egg! A risk worth taking, perhaps? But look at it another way: suppose there are 5 million elderly and young vulnerable people in the country, and each eats one egg a day, then a thousand vulnerable people will put themselves at risk each day.

The advice first presented to the public by the UK government and US Food and Drugs Administration during 1988 still stands. **Do not eat any raw egg.** This, to me, includes the yolk which is raw unless it is solid all the way through. Boiled eggs need 7–10 minutes; fried or griddled eggs should be cooked on both sides until all the yolk is firm. Scrambled eggs should be firm and omelettes cooked until no runny egg is left. Similarly, soufflés and baked eggs should be firm.

The egg producers and their advisers are not happy with the claim that a great deal of salmonella is caused by eggs. They point out the following:

1 **The number of occasions in which salmonella in a person has been proven to have come from an egg are very few indeed.** This is true, but surely a sad reflection on the failure to label eggs where and when they were laid, and that once eaten, there will be difficulty in testing the egg!

2 **The number of bacteria inside an egg will be small**

because of the lack of oxygen. Therefore there will not be enough of the bacteria to generate an infection. But we do not *know* the number of salmonella in an egg is small, nor how many salmonella are required to initiate food poisoning from eggs. Indeed, very few bacteria were dangerous in chocolate, another fatty food (Chapter 5).

3   **Most salmonella in food poisoning is caused by lapses of hygiene.** But if salmonella was not present at all in food, it could not multiply. Also, can the scale of the increase of salmonella food poisoning be blamed on worsening hygiene? No, it can't. And why is it that just one type of salmonella – *Salmonella enteritidis* – has increased so much since 1985 and not the other types? And why is it that *S. enteritidis* almost exclusively contaminates laying chickens and broilers? No: I am afraid, egg men, that it is your product to blame for most of the increase in salmonella food poisoning.

Campylobacter

This is the name for the commonest cause of food poisoning. In 1992 there were 35,000 reported cases of campylobacter food poisoning in the UK, with a real incidence thought to be 10-fold that figure. This slightly exaggerates its importance, since death from campylobacter is rare.

The campylobacter bacterium is shaped like a tiny comma or short spiral and is caught by eating contaminated poultry that is not adequately cooked, or from unpasteurised milk or even water. As with salmonella, campylobacter can be transferred to other foods following contamination by raw food, particularly poultry. Small children aged less than one year and adults in their twenties seem most vulnerable. Undercooked food eaten on camping holidays or at barbecues may be responsible.

The illness usually begins between 2 and 10 days after the suspect food has been eaten. Recent research shows that bacteria of this sort can survive well in the acid conditions in the lining of the stomach, so it is possible that we can become ill after eating just a few of them.

The first effect of the illness is a feeling of wanting to vomit. Vague tummy aches and a fever of about 38.5 °C (100–101 °F) may last a day or so. Acute pains and diarrhoea follow. Features of the diarrhoea are that it can be watery and contain fairly bright red blood. Muscle pains and headache may occur. The illness lasts for about 5–10 days, with a further 1–2 weeks required to return to normal health.

A barbecue story

The sun's rays flickered through the trees onto the crowd around the barbecue. It was still warm: but for the gentle breeze it might have been too hot. Rabbits were scampering about in the field, and the melodic song of blackbirds, and thrushes nearby and wood pigeons further away added sparkle to the early evening rosy glow. It was idyllic.

The women had seized with enthusiasm the offer of the men to do the cooking. The sense of well-being was enhanced by the succession of wines. Everyone was at peace with each other, and with the world. The men were intent on impressing that they could cook and had placed whole chickens in addition to cut up pieces on the grate over the glowing charcoal. The breeze fanned the heat onto the chickens only sporadically, but the sun lit up the food and the smell produced marvellous anticipation.

Not that everyone seemed interested in eating. The barbecue had taken longer than expected to become hot and the interval allowed more wine to be drunk. No-one seemed to care much what they ate.

When the chickens began to char on the surfaces nearest the heat, the men cut them into pieces with a cleaver and everyone marvelled at this. They were now nearly at the end of a memorable evening and it was going to produce a lasting impression.

Within 5 days, all the barbecue party were distinctly unwell. All felt weak and shaky. Most had bellyache that came and went in spasms. Many had other symptoms of an acute infection of the gut. The only person to escape was the chief barbecuer, who did not have time to eat any of the food!

The doctors were called and arranged for tests to be done at the local laboratory. After 2 more days, most people were still feeling ill, and the doctors told them that they were suffering from campylobacter infection. They were also told to expect a visit by Environmental Health Officers (health inspectors). When they came, they first tracked down what food was common to everyone. The barbecue. Was there anything odd about the food? How had it been kept before cooking? Where was it bought? Who cooked it?

The chief barbecuer had been happy to be able to go to work, but he was traced finally. He was emphatic that the chickens were cooked thoroughly (it had not occurred to him that it is almost impossible to cook a whole chicken thoroughly on a home barbecue). What about the cleaver used to cut the 'cooked' chicken up? 'Was it the same one used to cut the raw chicken?' asked the inspector. 'Well, yes' said the barbecuer.

Viruses

Viruses are the smallest agents of infection known and consist of a core of nucleic acid, the hereditary material, and a coat of proteins and sometimes other substances. The importance of viruses that may be eaten is that they can cause a disease very similar to bacterial food poisoning. The difference between viruses and many bacteria is that viruses just contaminate food, but do not grow in it, as bacteria can. In children, a small round virus causes vomiting, diarrhoea and abdominal pains.

Hepatitis is caused by a virus, and can be transmitted from contaminated food. The virus gets into food in two ways: raw or partly cooked fish, such as shellfish, may be contaminated by human sewage; and almost any food prepared by someone carrying the virus in his intestines may be contaminated.

E. coli 0157 – sandwich fever?

There is a new type of infection caught from food that has not been given a name by the media. It was first recognised in Canada. In 1985, 73 residents and staff in a nursing home developed severe gastroenteritis after eating sandwiches. All appeared to get better initially, but then 11 residents relapsed with kidney failure and died. The cause of the outbreak was found to be a bacterium called *Escherichia coli* 0157. This is normally found in cows' and other animals' intestines and gets into meat in the slaughterhouse.

Although similar bacteria exist in human intestines, they do not usually produce the poison responsible for the illness. The worrying aspect of this poison is that it is due to a tiny virus actually inside the bacteria, and it is possible that this virus might infect other types of *E. coli* and cause them to produce the toxin.

The disease in people is on the increase, now being the commonest cause of kidney failure in North American children. Officially reported figures for these bacteria in the UK are as follows: 1982 (1); 1983 (6); 1984 (9); 1985 (52); 1986 (77); 1987 (96);

1988 (86); 1989 (178); 1990 (382). The actual figures must be much higher and there were 181 cases of kidney failure due to this disease in 1989 alone.

Why are these *E. coli* 0157 bacteria and their viruses on the increase? It is true that in many instances it has been impossible to prove where the infection in any one person has come from, but food (or possibly water) must be the likely culprit. After all, the only likely route whereby a bacterium in a cow's guts can get into ours is by eating it.

What foods might be the cause? Unpasteurised milk is certainly possible, but it would not explain the rise in incidence. Sausages, burgers and any processed meats are more likely yet again as the cause of the problem, particularly if stored for prolonged periods at 8 °C as UK regulations allow. 'Yet again', because this also explains the cause of other diseases such as listeriosis. The guilty foods are likely to comprise either raw food, such as sausages not properly cooked, or cooked processed meats, subsequently contaminated and then eaten, as in sandwiches.

We have performed experiments showing that *E. coli* can grow happily at temperatures as low as 8 °C. More bad news for cook–chill!

Incidentally, what are we going to call this disease? How about Sandwich fever?

Botulism

Botulism is the classic food disaster, and should be entirely avoid-able. In the absence of air, and in the presence of moisture and nutrients, and not too low a temperature, the tiny dormant spores of the bacterium *Clostridium botulinum* start growing. These spores are found in the soil, and on some fish and vegetables, but as spores they are harmless. When they grow, however, they convert to bacteria and produce the most potent poison ever identified. This might occur in cans that have not been sterilised properly, in meat pies left at room temperature (as still happens from time to time in the USA), or possibly in any food where these conditions are right.

Food infected by botulinum toxin tastes all right, but the consequences are diabolical. After 12–18 hours, the toxin gets into the body and settles between nerves and muscles, preventing the nerves from activating the muscles with electrical impulses. The main effect on the patient is therefore paralysis. There may be difficulty in talking and swallowing, double vision, vomiting, staggering, shaking, inability to stand, collapse, inability to breathe and even death. On admission to hospital and careful treatment on life support systems, most people now survive.

But the real horror of the disease is that while the person affected experiences rapidly progressive paralysis, he or she is fully alert.

Food poisoning in the USA

During the last 10 years, the number of food poisoning cases reported in the USA has been much lower than in the UK, although the trend is also upwards in the USA. Salmonella reports have usually been in the low thousands annually, campylobacter and shigella in the hundreds or low thousands. It is only botulism that is typically higher in the USA than in the UK.

One reason for the lower USA figures is the difference in styles of medicine between the USA and the UK, with many more food poisoning cases missed in the USA. For this reason it is customary to multiply the USA figures by about 100 to arrive at the real numbers, and UK data by 10. Even allowing for this discrepancy, the chance of getting food poisoning would seem to be much greater in the UK than in the USA.

A few USA statistics may be of interest:

Year	Event in the USA
1985	At least 16,000 cases of salmonella from contaminated milk after pasteurisation in Ohio

1989–90 4288 cases of water-borne infection
1990   245 people develop salmonella poisoning
         from eating cantaloup melons[a]
         174 people develop salmonella poisoning
         from tomatoes[a]
         6 people develop paralytic poisoning from
         mussels in Massachusetts
1991   3 cases of cholera acquired from imported
         frozen coconut milk
1992   *E. coli* 0157 (Sandwich fever) found in most
         states, especially in the North

[a] Fruit and vegetable sources of salmonella are newsworthy because they are unusual, and are likely to be secondary to mammal or bird contamination.

Other food poisonings

The key to understanding *Listeria monocytogenes* is that this bacterium is fairly resistant to heating, so cooking may only stun it rather than eliminate it from the food. If cooked food is then kept cold – even at temperatures as low as 0 °C – listeria can recover after a few days and then multiply. When growing at low temperatures, *L. monocytogenes* produces an abundance of a toxin that reduces the ability of human cells to dissolve the bacteria. Fortunately most people have fairly good natural resistance to *L. monocytogenes*, since once it gets a 'foot hold', the resultant meningitis and/or septicaemia can be severe, with the risk of spread to the baby in the pregnant woman.

Food poisoning due to *Bacillus cereus* is relatively straightforward, resulting from boiled rice becoming contaminated with spores that germinate during too long holding at room temperature. The unsuspecting customer of the take-away may think that his fried rice has been freshly cooked from scratch, whereas a portion of the boiled rice will have been warmed briefly in a frying pan!

One bacterium, called *Clostridium perfringens*, causes diarrhoea and grows only in the absence of oxygen in the food, but needs

the presence of moisture and nutrients. Stews left at warm temperatures for too long are the main culprit. Staphylococcal poisoning results from the chance inoculation of food with strains that
just happen to produce a powerful poison, described as an enterotoxin, that does in fact act on the brain to cause vomiting. The
main medical features are shown in Table 2.

Most food poisoning bacteria and their toxins have no taste or
smell. The smell of putrefaction is usually due to relatively harmless pseudomonads. Hence we cannot identify whether a particular food is contaminated on the point of eating. The science of
food hygiene is therefore calculated to avoid the presence of these
contaminants.

Management of food poisoning

Ideally, all patients with gastrointestinal and other symptoms that
might be attributable to food poisoning should be investigated
thoroughly. This, we know, requires more resources than are
often available.

The actual approach must be aimed at

1   identifying the scale of the problem of food poisoning;
2   making specific diagnoses in individual patients; and
3   initiating investigation into the identity of contaminated
    food.

The major outbreaks are fortunately rare. With the presentation of several patients simultaneously with suspected food
poisoning, the doctor must refer faecal samples, possible vomit
and all the available food to the laboratory. Notification of the
incident to the appropriate officers is essential, and will initiate
a probe into the circumstances. It is important that the patients
do not warn a suspected caterer of imminent scrutiny!

The question of when to investigate sporadic (i.e. one-off)
cases is much harder, and depends on local facilities.

Although water is not strictly a food, we must be alerted to
the possibility of epidemics of diarrhoea caused by the protozoon

Table 2. *Food poisoning bacteria and their diseases*

Agent	Natural source	Other food or drink vectors
Salmonella spp.	Chickens, eggs, ducks, turkeys	Many prepared foods, notably cooked meats, bean sprouts
Campylobacter spp.	Poultry meat, unpasteurised milk	Water
Listeria monocytogenes[b]	Soil, animal manure, water	Chilled and processed foods, soft crusted cheeses, cooked chicken, pâté, salamis, recipe (cook–chill) meals, also salads
E. coli	Cattle	Burgers, sausages
Bacillus cereus	Widely in environment	Take-away fried rice
Clostridium botulinum[b]	Soil, vegetables, fish	Defective canned items of vegetables or fish, processed meat and vegetables
Clostridium perfringens	Intestines of mammals, including man	Stews, minces, anaerobic food
Staphylococcus aureus	Noses, groins, intestines of mammals including man, birds	Cream, cooked meats, custards, moist, processed foods

Incubation period	Typical patient	Symptoms and signs	Estimated number of cases 1991 UK[a]	Estimated number of deaths
18 h–2 days	Anyone, especially the very young and old	Diarrhoea, abdominal pains	300,000	100
5–7 days	Particularly young healthy adults. Immunity can develop	Bloody diarrhoea, abdominal pain, exhaustion	350,000	1
5 days–5 weeks	Pregnant women, elderly, immuno-compromised, also some young healthy adults	Flu-like illness in pregnancy, stillbirth, septicaemia, meningitis Few gut symptoms	400	100
? days	Children	Enteritis, kidney failure	500	50
6–18 hours	Anyone	Diarrhoea, abdominal pain	1000	0
12–18 hours	Anyone	Progressive paralysis of motor nerves	27	1
12–24 hours	Anyone	Diarrhoea, abdominal pain	5000	0
1–3 hours	Anyone	Vomiting, sometimes severe, with haematemesis	5000	5

[a]Based on the official number of cases reported up to end September to the Government's Communicable Disease Surveillance Centre, with multiples (e.g. ×10 for *Salmonella*, ×3 for *Listeria*) to allow for inefficient reporting.
[b]Figures for 1989.

parasite, cryptosporidium. This is well known to occur in AIDS patients, but may also occur in otherwise healthy patients, especially children. Although there is no treatment, it is worth attempting to identify the parasite in faeces. This enables the outbreak to be traced back to defective sand filtering from reservoirs.

Treatment

Most food poisoning results in loss of water and salts, with a risk of possible complications. Fruit juices are the mainstay of treatment. Antibiotics have little role, even against campylobacter. Although this bacterium is destroyed by the antibiotic Erythromycin in laboratory experiments, it is not helpful in practice because the disease clears up spontaneously in time and clinical trials have not shown benefit from Erythromycin. Antidiarrhoeal drugs such as codeine and loperamide may be helpful. Most patients can be looked after at home. However, the 1–2% of patients with the spread of salmonella into their blood may require treatment in hospital for septicaemia and abscesses. Antibiotics for these include trimethoprim, chloramphenicol, and ampicillin. Quinolones, such as ciprofloxacin, may be effective but should be used sparingly because of the risk of selection of resistance.

The number of people who contain the salmonella bacterium in their intestines months after the acute illness must have increased recently, and there is usually no point in repeated testing of samples to see if it is still there! People generally return to normal activities following the disappearance of the diarrhoea, with instructions for meticulous personal hygiene, i.e. hand washing. A few patients who work in health care or catering may need specific management – it is suggested that this is discussed with their employers and, if necessary, the appropriate Medical Officers. Even with these people there is a trend towards a less aggressive attitude over staying away from work until the salmonella has disappeared. There is no proof yet that some of the new antibiotics might reduce the length of time people carry salmonella in their guts.

Listeriosis

The management of patients with suspected listeriosis, particularly in pregnancy, depends principally on taking samples of blood. *Listeria monocytogenes* grows readily in the laboratory. The disease can mimic influenza. Although the disease due to this bacterium, listeriosis, is not common, proving a link with any particular food is difficult because of the long interval between eating the food and becoming ill.

BSE

Bovine Spongiform Encephalopathy (BSE), or Mad Cow Disease, is a terrifying and unpredictable disease both for cattle and also for us. Arguments over it have raged in the UK for several years. The central facts are very few indeed. The comments, opinions and advice have been prolific. So what are the facts?

The first is that the cause of the disease in cattle is a tiny, very tough (that means it is not killed by cooking) type of 'virus' that is similar to those that cause infections in sheep (the disease scrapie), goats, deer and man. Our diseases are the uncommon dementia, Creutzfeldt–Jakob disease, and kuru in the cannibalistic Fore tribe of New Guinea. The next fact is that the interval between the virus getting into the body (often by eating it) and the final illness is long: from about 1–2 years in a small animal to an estimated 5–30 years in people.

What is not known is the exact composition of the agent that causes the disease. There are two theories. One, essentially the British view, suggests it is a type of virus with some DNA within its protein covering. The other, the view of American researcher Stanley Prusiner, is that it is a tiny hardened protein without any DNA, called a prion. Because we do not know for certain what the infectious agent is, we must be careful about any claims for safety, or indeed danger.

The following summarises the main milestones of the BSE epidemic in the UK.

1985/6 First few cases of BSE confirmed as a result of sponge-like appearance (under the microscope) of the brains of cows dying after a staggering illness lasting weeks or months.

1987   Around 400 cases of BSE during the year from widely separated areas of UK.

1988   Cases continue to rise to over 2000 in the year. Seriousness of the problem appreciated by UK government, and the likely role of offal from the rendering plants being the cause. Feeding of cattle, sheep, goats and deer (all ruminants) with offal products banned on 18 July 1988. Government appoints committee, chaired by Sir Richard Southwood, Professor of Zoology at Oxford to look at causes, likely future progress of the disease and action needed. Media interest moderate. Beef-eating and slaughter prices of carcasses unaffected.

1989   Cases now exceed 7000 for the year. Southwood reports in February 1989, stating that the cause of BSE was the rendering of scrapie-infected sheep brains and the subsequent inclusion of the infected material into cattle feed. Considers that cattle will be a dead-end host and that the disease will not spread once the source of it has been cut off. Southwood goes further: he makes a bold prediction that cases will peak at 350–400 per month and the total will be 17,000–20,000 before the disease dies out in the early 1990s. Government reassurance is based on the premise that BSE *is* sheep scrapie and that there is no reason to believe we can catch scrapie. Media suspicious. Television programmes begin to question whether BSE can pass from cow to calf (vertical transmission). Beef consumption still unaffected.

1990   Total BSE cases for this year, now 14,000. Southwood's predictions overtaken by events. Media interest intensifies, culminating in May 1990 with public boycott of beef and refusal of many countries to accept British beef. In July 1990, Parliamentary Agriculture Committee (members are MPs with farming interests) attempts to reassure the public that beef is safe, by mounting personal attacks on those raising doubts. Consumption of beef slowly rises. The Meat and Livestock Commission is successful in promoting beef to schools. But the auction price of beef does not return to previous levels. The surplus is bought by the government and stored deep-frozen. This intervention buying gives a false impression of the volume of beef consumed. Farming land values begin to fall. The compensation paid to farmers with BSE animals is increased from 50% (since 1988) to 100%. Various cattle organs, such as brains, spleen, and thymus, have been excluded from the food chain, but all the edible organs of calves under 6 months of age still enter the food chain, and animals over 6 months – even from infected herds – are exported. Commercial damage to the beef industry is minimised. Zoo animals such as elands, antelopes, oryx and a panda, and also domestic cats, are described as suffering from a 'BSE-like' illness.

1991   The total BSE cases for this year is now 25,000. The government's advisers now predict that the numbers will start dropping in 1992, and that the disease was caused by the re-feeding of offal derived from cattle in addition to sheep. No research is being undertaken to identify how many cows are infected but not yet ill.

1992   By June the notified BSE cases are running at 1000 weekly (suggesting a total for 1992 of 40,000). The government still believes that numbers will begin to fall by the end of 1992. The government is required, through formal parliamentary questions, to reveal details of animal experiments. Of experiments completed, seven species of mammal have been challenged with BSE. These are mice, goats, sheep, cattle, pigs, marmoset monkeys and hamsters. Some of each of these species, hamsters apart, have gone down with BSE. In the summer of 1992, cases of BSE are 10-fold their predicted numbers, despite the offal ban 4 years previously. Evidence for vertical transmission accrues.

Creutzfeldt–Jakob disease: where does it come from?

Officially notified cases of Creutzfeldt–Jakob disease (CJD) are of the order of 50–60 annually in the UK The real range might be as high as 1500–9000 because many people with dementing illness are missed. If BSE did infect man, it is probable that the resultant disease would be CJD.

The study of BSE has produced overwhelming evidence of the existence of an infective agent acquired through eating. Thus this group of agents as a whole should be considered as being acquired through eating food. CJD is therefore probably acquired from meat (if you don't like this suggestion, try and find an alternative). The three major sources of animal meat throughout the world must be scrutinised very carefully: i.e. products from cattle, sheep and pigs. The possibility that sheep scrapie might be responsible for CJD has been considered by several workers. Because the incidence of scrapie in sheep varies from one country to another, being highest in Iceland and lowest in New Zealand it has been possible to look at the varying incidences of the two disease in a number of countries. No association has been found. Sheep scrapie, as the UK government maintains, is of little direct danger to us.

Pigs might be responsible for CJD, but a number of factors argue against this. First, pig meat is generally eaten when the animal is 5 or 6 months old: that is, at an age when it would be expected not to be highly infectious. Also, a high incidence of CJD occurs in Libyan Jews, whose diet presumably excludes pig products.

This means that, with other sources excluded, cattle must be the most likely source of CJD. Consistent with this proposal is the frequency with which cows are slaughtered at the end of their lactation (typically between 6 and 7 years). Being elderly must be a highly significant factor in the availability of the infectious agent because it is known that the older animals become, the more infection they carry in their bodies. If cows are the source of CJD, then this means that the infection was acquired many years previously. If the incubation period of CJD is 5–30 years, then few current CJD cases can be caused by the BSE agent in the 1980s. It is certainly possible that a BSE-like infectious agent could have been present in cattle for many years either without producing clinical infection, or indeed responsible for occasional clinical disease. There are claims of the existence of cattle suffering a BSE-type disease many years ago in Yorkshire, UK, or more recently in other countries, and such events would be compatible with this view.

If many of the claimed 1500–9000 cases of CJD in UK annually have indeed been caused by consumption of beef products some years ago, the potential danger from BSE since 1986 becomes disturbingly obvious. The most worrying hypothesis does suppose that the infectious agent from BSE-infected cattle is responsible for CJD. If this is the case, the high incidence of infected animals must present a phenomenal danger to man from around the turn of this century into the next.

10
Additives

Total confusion now reigns. Are additives dangerous or not? Does the long list of chemicals put into food deter you from buying it? Do they have any meaning for you? What, for example, is the significance of propane-1,2-diolalginate or hexamethylenetetramine? Well, why are there so many chemicals to put into food?

The unpalatable and simple answer is that you want them. Before you become too indignant and start blaming the food industry for acting for its own convenience and profit, let me explain.

Additives are in all manner of food you buy and enjoy: ice cream, sweets, biscuits (cookies), soft drinks, yogurt, beer, cakes, smoked fish, cheese, soups, instant potato, bread, crisps (chips), cakes, sausages, chewing gum, wine, beefburgers, frozen chicken and so on, and so on.

Many processed food items can be available only because of additives – either to ensure the product lasts sufficiently long for you to purchase it at your convenience, or to provide a pleasant taste, colour or texture, or to prevent it from going bad. You would simply be unable to obtain many of these items without additives. This is the central argument of the food industry. It says that it only produces what you want and the packaging now tells you the names of all the chemicals put into the food, and if you do not understand what the names mean, that is your problem. You always have the option of not purchasing it. Free society, free choice and all that!

The food industry is, of course, aware that it has to make the product as appealing as it can, so it does use chemicals to make its products irresistible to you. Packaging, advertising, display and other types of promotion are all employed for greatest effect. The resultant products may well be completely unnatural (in which case they may be referred to as natural), stale (described as fresh), free from artificial colours (full of artificial flavours), new (newly promoted), improved (more expensive) or old-fashioned (modern mass-produced).

Many of you have been quite happy to be seduced by this approach. The food industry knows what you like, how to influence your children and how to sell it to you. In the final analysis, you are responsible for buying and eating additives.

This brings us to the first concern over food additives. Once their use in a particular product has become established, the cost of the mass-produced, additive-enriched product can drop because of the huge volume of turnover. The low-volume, additive-free items may be more expensive. This may further distort the market, with whole or additive-free food confined to the few high-price specialist retailers.

OK, you are thinking. 'You've explained why we eat so many additives, but I have bought this book and I want to know whether the additives are safe for me or my children. Are you going to let on?' Yes, I will try but the issue is not quite as simple as it sounds. We frequently have to say – on the evidence available at present – that additives are probably safe as long as you do not eat too much of them, or you are not pregnant, and so on.

We are now faced with two problems. First came the awareness that some chemicals might be dangerous in food, after they had been eaten for years. It is only in the last century that we have appreciated that the following can be dangerous: exposure to X-rays and UV light, smoking, asbestos dust, thalidomide, DDT, CFCs, Opren, chloramphenicol, to name a few.

The second dilemma concerns which procedures should be used in order to establish safety. It is no good demanding that someone proves safety unless we know what experiments should be done and what they mean. To start with, there are experiments that we can't do: we can't deliberately experiment on people. Doctors and scientists do try to study associations of certain diseases with eating food additives. This takes decades and the results can be impossible to analyse. Suppose we tried to look for an association between any one disease and any one additive. Let us, for example (quite hypothetically), assume that we found an association between fits (or epilepsy) and eating the brown colour caramel. To prove cause and effect between the additive and the disease – that is, the caramel caused fits – we would have to show that other additives were not responsible or that people who developed fits tended to eat more caramel than people who didn't suffer from epilepsy. The final proof of the link between cigarette smoking and cancer was to show that cutting down smoking reduced cancer. Well, why not see if cutting down caramel stops fits? No good, because anyone who develops fits is given long-term drugs as a precaution against more fits. Would it be ethical to stop this treatment? No. This does mean that any allegation that a food additive causes a disease is extremely difficult to refute or to prove. And this can be interpreted as *carte blanche* for the food industry to poison us, or as a scaremongers' charter!

So if we can't experiment on people, surely we can do laboratory tests using either human cells or bacteria or experimental animals. The problem is that many different types of experiment need to be done as no single one reflects accurately the effects of the additive on people. Then there is the growing concern about cruelty to laboratory animals. It really is true that animal experiments may not predict closely the effects of a chemical in man. So the thorough testing of any one chemical additive takes years and millions of dollars.

Put another way, we just do not possess the facilities to test all the additives in current use comprehensively, and in any case there is no test system guaranteed to give an accurate prediction of the long-term effects on people or their children. There is growing awareness that damage to the DNA of male sperm or female ova could result in disease such as leukaemia in children born subsequently. How can this be researched for any one chemical?

However, there are reasons to believe that certain additives are not dangerous at all, or perhaps minimally so, for the following reasons.

1 Some have been in use for sufficient time without any known evidence of danger. The preservatives acetic acid (from vinegar) and sorbic acid are examples.

2 Some substances are identical to natural products about which there are few worries. For example, caramel is produced whenever bread, pastry or cakes are baked. However, it cannot be assumed that all natural substances are safe – poisonous natural fungi and berries are still poisons. But a 'natural' substance has usually been eaten for centuries, so we have good experience of it; if there is a potential danger we should have had an inkling.

3 Some substances are so similar to known safe products that safety can also be anticipated, but this does not actually prove safety.

What might be the risk from untested and unresearched additives? The food industry and the UK Ministry of Agriculture (the two are not identical) have been quick to point out that there is no proof of any serious ill effect from any chemical. They still defend the availability of the yellow dye tartrazine on the grounds that the alleged overactive behaviour in children attributed to it has not been proven.

To obtain proof, we have seen, is very difficult under any situation, and over the short term quite impossible in any one individual. Children might be overactive even if they don't eat tetrazine. Some, I know, are real devils! Consider a heavy smoker for 40 years who dies of lung cancer – it is impossible to establish a proven cause and effect in this person because it is a distinct reality that he might have succumbed from lung cancer (or a heart attack, for that matter) *even if he had not smoked.* We know that smoking is dangerous because of the prevalence of various diseases in *populations* of people, not in individuals.

Take additive X. Doubts over it are raised by what the food industry describes as 'food terrorists' – that is, academics outside their influence whose comments might put their profits at risk. The industry will wheel out one of its paid consultants, who will say that the chance of any toxicity (damaging effect on a person) is less than one in a million. Everyone is reassured – we continue to eat ice creams made in the image of a yellow Empire State Building. But suppose the risk of ill health from that additive is a little greater than 1 in a million: say, one in 100,000. This might not be too disturbing. But wait a minute. How many additives are there? Perhaps 10,000. Suppose this type of risk applied to all of these, so the combined risk could be as low as 1 in 10. This is the problem: the cumulative effect of eating so many chemicals might be considerable and many have not been used for long enough to exclude toxic effects.

Is this pessimism justified? Surely we have a right to enjoy our coloured, flavoured, sticky goodies? I just wonder whether 'food terrorists' might have overdone the risk from additives, because it is the easiest aspect of processed food on which to focus. Perhaps instinctively 'they' don't like the environmental and social implications of so much processed convenience food. I wonder. Am I a food terrorist?

What are additives for?

First, to make food look attractive. Much of our appetite for food is based on sight. There is a nerve pathway from our eyes to the brain and then to the stomach, that both causes hunger on seeing something that looks appealing to eat, and prepares our stomach by secretion of digestive juices. It is not just us. Blackbirds are particularly partial to red berries, such as my home-grown raspberries. The main colours used in food are listed below.

Colour added	Chemical used
Brown-black	Burnt sugars (caramel) Carbon black
Yellow	Carotenoids Riboflavin Quinoline yellow, sunset yellow Tartrazine

Green	Chlorophyll
	Copper compounds of
	chlorophyll
Blue	Brilliant blue
	Anthocyanines
Red	Erythrosine
	Red 2G
	Cochineal
	Beetroot red
Purple	Methyl violet (suspended)
	Violaxanthin

Critics of food colourants point to their lack of nutrient value and claim they are unnecessary. Of course, these claims are valid, but we are hooked on them. We have come to expect that certain products should be a certain colour and won't eat them unless they are.

One potential hazard should be mentioned. Crystal (or methyl) violet, which is a purple dye for staining bacteria in laboratories and also for treating some infections, such as in the ear, has been found to induce damage to DNA (i.e. mutations), so it is theoretically capable of causing cancer. This has not been proven in people, but its use for labelling the skin of oranges has been discontinued because of this risk.

Preservatives

We need preservatives in food if we want instantly and reliably available perishable food that would otherwise decompose.

Agent	Foods used
Benzoic acid, benzoates	Soft drinks, beer, fish, fruit pulp
Sorbic acid, sorbate	Low fat margarine, processed

	cheese, soft drinks
Sulphur dioxide, sulphites	Sausages, beer, wine, cider, biscuits, potatoes, fruits
Hexamine	Herring, mackerel
Propionic acid/ propionate	Bread, cakes, Christmas pudding
Nitrites, nitrate	Ham, bacon, corned beef, sausages, pizzas

The principal concerns in this list are nitrates (and nitrites) and benzoic acid. There is a possibility that these, at least in high doses over long periods, can induce cancer, and it is interesting to note that the water authorities are endeavouring to reduce the amount of nitrite and nitrate in drinking water.

Artificial sweeteners

Artificial sweeteners have become popular because many of us know we are eating too much sugar – no doubt partly due to clever advertising on television. There is nothing wrong with sugar itself; the problem comes from eating too much of it.

So, what are the alternatives? Saccharin is one of the best known and is, by weight, 300 times as potent as a sweetener compared with sugar. It is an artificial chemical derived from coal tar and was first used in 1900. With saccharin there are two problems. One is that the sweet sensation can change to a bitter aftertaste. This does not happen with sugar. The other is that high doses of it can cause tumours in mice. It is probably advisable not to exceed 2.5 mg per day (if you can work it out!) as laid down by the World Health Organisation and the European Community.

Then, there is aspartame. This is 100 times as sweet as sugar and is made from the chemical union of two amino acids, the building blocks of protein. Unfortunately, one of these amino acids is phenylalanine, which makes the substance dangerous for the relatively small number of people with the genetic disease, phenylketonuria. There are also claims and counter-claims over whether it can cause nervous disorders. Aspartame is very commonly used in diet drinks, and some of the alleged side effects of aspartame could be due to other ingredients such as caffeine. The official view (i.e. of our governments) is that, if eaten in moderate amounts (but what, you ask, does that mean, and I don't know), the side effects, if any, should be minimal.

Cyclamate is a simple synthetic chemical that has been blamed for toxicity. But others claim these fears were caused by experiments done in mice that were fed large amounts of both saccharin and cyclamate, so it was not apparent which one causes the tumours. Cyclamate may become generally available in the EC very soon.

So, if we want sweet drinks, we seem to have the choice of eating too much sugar or synthetic chemicals of disputed safety. One way to resolve this problem is to educate our tongue (where sweetness is sensed), from early childhood, to resist over-sweet tastes. Is the answer tap water?

But the main problem with food is that so much of it, from cereals and chocolate to baked beans, contains added sugar, partly to counter the effect of salt.

As the advertisements tell us, sugar is natural and quite acceptable in fruit. But this does not justify the extent of its use in processed food.

Other additives

Antioxidants prevent oxygen in the air chemically reacting with, and then decomposing, some of the nutrients, such as polyunsaturated fats. Artificial chemicals such as those called gallates can be used, but so too are vitamins C (ascorbate) and E (tocopherol). These are put into bread, meats, soups, drinks and other foods.

Emulsifiers and stabilisers keep the texture as desired, and include natural gums and cellulose, detergent chemicals, and fats. The need for so many flavours to be added alerts us to the loss of taste caused by processing. These flavourings can be 'natural' in the sense that they are extracted from plants; for example, vanilla essence from Central American vines. Some flavours are completely artificial, often synthesised from products derived from coal tar.

The exact identification of an artificial flavouring is rarely given. The food industry argues that it might be giving away a trade secret, and in any case the amount of the chemical present is very small.

But people can become 'hooked' on certain flavours, and could over the years eat a great deal of a particular flavouring chemical.

Food intolerance

The principal worry over many additives has been the possibility of long-term serious effects such as cancer.

But many people – and it really is many – experience fairly immediate symptoms after eating processed food. The symptoms include a feeling of sickness (nausea), actual vomiting, a bloated sensation in the abdomen, colic and diarrhoea. Skin rashes and headache are common. Sometimes it is easy to identify the culprit food; sometimes it requires meticulous analysis. No wonder one of the fastest growing medical specialities is in food intolerance.

The food industry has of course learnt to defend their additives, as shown by the following telephone call.

A true telephone call

'It's the Monosodium Glutamate Information Service', announced the young lady down the telephone.

'Yes', I said in disbelief.

She continued enthusiastically, 'We represent the interests of the producers of monosodium glutamate, and you have been saying unpleasant things about our industry.'

'No', I protested, 'there is nothing wrong with monosodium glutamate; it's just the way it's used that is the problem.'

'What do you mean?', came the reply.

I tried to explain. 'My colleague, Dr Stephen Dealler, has found that if you put monosodium glutamate, or even ordinary salt for that matter, into convenience meals and then microwave them, the waves react with these chemicals so they can't penetrate into the centre of the food. This means the outside gets hot but the inside stays cold.'

A long silence was followed by, 'What do you suggest?'

'If you put flavour enhancers and salt into composite foods, don't use a microwave', I ventured.

Trying to divert the blame, she queried, 'What do the manufacturers of the food and the makers of the microwave ovens think of this?'

'Either they say it isn't true, or if it's true, it does not matter, and in any case, this has been known for 20 years!' I replied.

Confusion.

This was a true telephone conversation I had recently and it highlights the extent to which commercial activities employ public relations fronts that protect any member having to defend its product publicly. Try asking a question of the Minister of Agriculture or the Department of Health in the UK or equivalent bodies in the USA, and you will be rewarded by 'A spokesman' or, more often, 'A spokeswoman' evidently selected for the art of evasion through the telephone, rather than being able or intent on answering any particular query.

For example, the 'Food Safety Advisory Centre' was set up in 1989 to represent six UK supermarket chains. Most people are not aware of the extent of these organisations. To get at the truth, I advise the media to insist on any spokesman defending a problem being named, and that person's status and experience clarified.

A few years ago, the 'Chicken Information Service' made the possible mistake of inviting me to a promotional presentation on chickens in front of the chief guest, Mr Donald Thompson (now Sir Donald), then a junior minister in the Department of Agriculture. Soon after this, I heard that one of the Sunday newspapers had reported an alleged comment. After being asked by a reporter why he wasn't eating chicken on the occasion, Donald's reply was, 'I'm minister of food, so I know what's in it!'

Conclusion

I am going to end this brief comment on food additives by suggesting that the main objection to them is not actually on grounds of health, but that their availability results in environmental concerns and a decline in the quality of our basic food. It is the need for so many additives, often resulting in simulated or faked food, that is the issue. The extent of food processing and the scale of additive use does give me cause for concern.

PART IV
Getting ready to eat

11
The supermarket

Rubbish (trash)

At long last, we are beginning to be concerned about the amount of rubbish generated by the supermarket culture. But why so much packaging? The pressures have come from both the sellers and the buyers. The seller needs the container to hold portions of the increasing amount of processed food. This accounts for 70–80% of all food purchased in the USA and 60–70% in the UK. For example, you might be able to transfer an unwrapped whole lettuce to your bag or box, but you would have problems with coleslaw!

Then the retailer uses wrapping to provide space for pictures, advertising and, indeed, details of composition and chemicals added in processed food, usually now required by law. It is always amusing to see how the pictures glamourise the actual product. And, of course, the packaging can sometimes disguise the real nature of the product. The retailer has learnt how important the image of the product is to the consumer. Much of the enjoyment of eating is associated with the imagination and the emotions: clever packaging reinforces the image, as does advertising, with many products angled at a particular age or socio-economic group.

The customer needs the packaging, too. Because he or she often wants to do all the weekly or monthly shopping in one speedy visit, the individual products must be prevented from spilling onto others. For example, it would not do at all for meat juices to get onto cheese, for reasons both of appearance and of safety. In the process of loading up the trolley, packaging has to be sufficiently robust to prevent damage to the items underneath the pile. The requirement for so much packaging is therefore caused by the way we shop, and our insatiable demand for processed food. It is not due just to the promotional strategy of the retailers.

Authorities have begun to respond to this state of affairs by encouraging or even requiring recycling of packaging materials. Many individual supermarkets have introduced their own schemes, and the packaging industry has developed biodegradable (i.e. not plastic) wrappings. A few consumers (hopefully you) have altered the way they shop in order to reduce the amount of trash generated.

But the amount of recycling varies from country to country. Holland, and especially Germany, are deep green in this respect. If you want to buy a 3 litre bottle of drink in Holland, it will almost certainly be in a glass bottle – good for weight-lifting training. I would call the USA pale green in this respect: there are some excellent voluntary recycling facilities. Meijer's scheme, of customers being paid 10 cents for each returned can or bottle, is to be applauded. In the UK, the colour is best described as yellow, with the government frightened of interfering with free market forces, particularly in its food industry with which it has had an incestuous relationship for years. However, UK consumers have begun to take the initiative and local authorities are attempting to separate rubbish with potentially recyclable components, such as cans or bottles.

But is any of this the real long-term answer? Would it be preferable not to create the rubbish in the first instance, so that there would be no need to recycle it at all? Do we really believe, for example, that our daily fluid needs should be met by drinking cans of sweet brown fizzy fluid, rather than water from taps?

The supermarket culture, like the take-aways (Chapter 13) and eating out, is part of our motor-car society. It is not the only way that human society can be structured. Local communities and local shops provide the alternative, but we may need to retrain our legs from pressing on motor-car controls to taking successive steps on the ground. This, incidentally, is technically called walking.

But for the time being, until the oil runs out, the supermarket culture seems to be here to stay and it is our responsibility to reduce the amount of packaging bought, and then recycle any we can. If I was in charge of the various economies, I would introduce a heavy tax on plastic packaging!

The really good buys in the supermarket

The scale of the supermarket chain's operations, with its bulk purchases and its incredible power to put downward pressure on the price of raw food, means low prices to the consumer. This may not be good news to the farmer, who now has very much to obey the demands of the retailer.

Substances like flour, sugar, butter, oils, margarine – the basic ingredients of much processed food – are good buys. But flour seems to be disappearing from supermarkets, particularly because people are doing less home cooking – isn't making pastry a bind? – and partly, I suspect, because the retailer would rather you purchased the manufactured (a word, incidentally, that means 'made by hand') product.

Cereals

All supermarkets should display an excellent range of cereals. The supermarket's own product may be equivalent or indeed *identical* to the branded goods, but it might be a cheap imitation. Whilst over-elaborate packing causes concern, at least the cardboard of most cereal packets is biodegradable. My favourites are Apple Cinnamon and Toasted Oats (available in the USA), and good old-fashioned unadulterated Kellogg's Cornflakes.

Fresh fish

There has been increasing incorporation of fresh fish stalls within the supermarket. Frequently an assistant will be available to skin, fillet and advise. Certainly this happens more frequently than with meat, which is usually ready packaged. I am certain that the future interest in fish will include the testing and adoption of some deep-sea species which are currently unfamiliar to most people. Fish does not have to be white – nor does it have to be bright orange. I don't like farmed salmon, but most supermarket deep-water fish is thoroughly recommended.

Canned food

Canned food, in the main, comes into the recommended category. In the main, because some developments in the canning industry are causing some disquiet. By the 1940s and 1950s the can had reached the end of its development to provide a rigid and safe container for its heat-preserved contents. This provided a store of food for months or years: not as good as fresh food, but definitely useful.

It was largely the availability of the microwave and the impossibility of heating food in metal cans that stimulated the use of plastic containers to hold 'canned' food. Some of these are very flimsy and their top may be sealed by foil. In theory these should be safe, but in practice any damage could break the seal or penetrate the container with introduction of dangerous bacteria. So please check the container before purchase and then store it carefully.

I think some of these microwavable containers quite amusing, such as the little bowls containing chopped chicken. The top is sealed with foil over which is a lid containing four holes. You simply remove the lid, peel off the foil, replace the lid, microwave for 2 minutes, stir and serve! Does this represent everything desirable and successful in our society, or everything rep-rehensible?

Another trend with cans is the use of the ring pull, so there is no need for all that hassle with the can-opener. First used on drinks, the ring pull has spread to canned sardines in particular (why?) and also to pet food. The junction where the lid is pulled away from the rest of the can has of necessity to be very thin, and there must be a danger of damage causing premature separation of the lid. So, again, checking is needed. I have actually experienced such a defect with dog food. The lid literally 'came away in my hand' after touching the ring. The side of the can had been knocked at some time and this must have damaged the seal. The top of the contents was covered in obvious fungus, and goodness knows how many other nasties.

Frozen food

Much traditional canned food has been replaced by frozen food, often quite justifiably (see Chapter 7). In the supermarket, it is increasingly difficult for consumers to distinguish frozen from chilled food, and some serious disasters could occur if frozen food was mistaken for chilled food and, for example, stored in too warm a refrigerator.

The traditional way of identifying frozen food is through the build-up of crystals and the presence of visible steam. Most supermarkets' deep-freeze cabinets contain a regular defrost (i.e. warming-up) cycle to prevent build-up of ice deposits, and the circulating air may be too dry to cause condensation of water vapour and hence will not produce visible steam.

The ideal deep-freeze container is either a floor standing cabinet with a lid or a wall cupboard with a door. Within those spaces the temperature can be carefully controlled. You may see temperature gauges, hopefully showing below $-18\ °C$, except in defrost cycle when you may mistake the food for chilled. Many supermarkets have computerised, out-of-sight monitoring of their deep-freeze temperatures.

Presumably because market research has shown that the effort of opening a door or a lid deters the prospective buyers, many deep freezers are now open. Floor standing cabinets with open lids (at least during shopping hours) must be safer than the deep-freeze shelves that I have seen in the USA, but not in the UK. But surely is not the waste of energy unacceptable with either open design? Energy is needed to cool down the air in the freezer and often warm up the ambient environment, but the cold air escapes into the room space, and the warm air there gets into the deep-freezer! What on earth is going on? Why not make doors and lids mandatory?

Chilled food

The same comments and problems are associated with chilled food, but the consequences of temperature abuse are more dire than with deep-frozen. If a deep-freeze cabinet is at −8 °C rather than −18 °C, then the food quality may deteriorate a little, but there will be no danger of bacterial growth. But if a chilled cabinet is at +13 °C rather than +3 °C, then serious dangers could develop during a day or so. This explains some of the concerns over the chilled shelves that seem to be used for almost everything, with the implicit propaganda that if it is cold, it's fresh!

Some products do not need to be kept particularly cold: these include standard butter and margarine (with less than 20% water), hard cheese, yogurt, and pork products preserved with sodium nitrite. These would be safe at +10 °C, certainly for a day or so.

But other products must be kept really cold. These include low-fat butter, margarines and spreads that do not contain preservatives, and processed meats without preservatives. There seems to be a major epidemic of pâté and what I call 'spam' type products in the USA. There is even pâté with lumps of cheese in it. Again, unless these contain preservatives, they must be kept below +3 °C. In passing, I saw one spam-type processed meat that was very pink, but I think it was beef-based, stating proudly, '80% fat-free'. What that actually meant was it contained 20% animal fat, of course, disguised by the processing.

Coleslaw and other chopped salads must be kept cold, as must cream desserts. But whole fruit and vegetables need not: certainly if stored too warm, vegetables wilt and discolour rapidly, but should not be dangerous. This means that foods are displayed on refrigerated shelves or in refrigerated cabinets for a variety of reasons and the open shelves without adequate temperature control are a botched compromise for convenience. I am sure you will hear more about this in the future.

For you the consumer, the following advice is offered for purchasing chilled and frozen food.

1 Prefer food stored in chests or cupboards with lids or doors rather than open shelves.
2 Check the temperatures of the gauges, if you can see them (-23 to -18 °C for frozen, 0 to $+3$ °C for chilled).
3 If there is likely to be a delay of more than one hour in transit, use cold containers or bags for transport.
4 Check your refrigerator and deep-freeze temperatures when opening the doors or lids.

Eggs

Eggs are probably the major cause of salmonella food poisoning in the UK, perhaps in the USA also, if not in the whole world (see Chapter 9). We are going to have sort eggs out in the future.

In the supermarket, you will see quite conflicting approaches to eggs, depending on whether you are in North America or the UK. In the USA the favourite colour of the shell is snow-white and the eggs are refrigerated. In the UK brown eggs are dominant – indeed you will be hard pressed to find a white one! In the UK the brown eggs are stored at room temperature.

The brown and white eggs do reflect slightly different breeding and laying stock, and it is not surprising that types of salmonella vary also. In the USA it is mainly *Salmonella enteritidis* type 8, and in UK it is *S. enteriditis* type 4.

In the UK the boxes of eggs frequently have slogans, such as 'country-fresh', names of farms, and pictures of the sun, meadows and trees. But most are battery laid. At present the European Commission is introducing controls over what can be legitimately called 'free-range' with different categories. The UK government does not seem enthusiastic over these proposals. There does not appear to be intended deception in the USA: the eggs are straight battery-laid.

The arguments for and against refrigeration go as follows. In favour of refrigeration is the belief that the cooler the egg is the slower it will dry out and deteriorate, and indeed prevent multiplication of salmonella, were any to be present in the egg. Against refrigeration is the claim that water might condense and drip onto the shell, and dissolve away the plugs conferring protection against entry of bacteria. Also, it has been argued that if refrigeration is needed to prevent growth of salmonella, then the chicken should be kept refrigerated because the egg is 2–3 days in the bird before it is laid! (This is not a serious suggestion.) Eggs which are cold are hard to cook thoroughly, particularly with frying or scrambling, so should be moved to room temperature an hour or so before cooking. But the average UK household is most unlikely to rise an hour earlier than usual to take the eggs out of the fridge!

In both the UK and the USA, eggs at the point of sale rarely state where and when they were laid. For a perishable, potentially contaminated product, this is quite disgraceful. The information usually identifies the packers and distributors only. The need for labelling as to exact origin is in the tracing of food poisoning. Once an egg is consumed it cannot be sampled, but the box may be recovered and the source could be identified. Heavily infected layers can then be treated appropriately.

At the time of writing, there would appear to be positive efforts to reduce salmonella from egg layers in the USA. The attitude of the British government has been depressingly inadequate and it is no surprise that the salmonella problem is greater in the UK than in the USA.

Sell-by, Best-by, Eat-by

These dates have usually been arrived at by research performed by the food industry on perishable food and primarily geared to the length of time the produce is palatable, rather than to nutrient content or the risks of bacterial growth. The dates are usually established using the anticipated temperatures and conditions of storage, with some latitude built in for abuse. This does mean that certain products might be edible a few days outside their sell-by dates, and it is not surprising that there is a flourishing trade in cut-price, out-of-date processed food discarded by some retailers. There are claims that some retailers replace the label after the failure to sell within the allotted time. Whilst these dates were initially intended as guides from the food industry, successful prosecutions have been achieved against retailers selling goods outside their sell-by dates. Despite this, the legal status of these dates must be in question, and there is an urgent need for agreed international standards for the times and temperatures for the safe holding of unpreserved processed chilled food.

Despite the imprecise means of arriving at these dates and their doubtful meaning because of the variable storage temperatures in the shop and in the home, the only practical advice I can offer is to keep within them. (I am not going to be tempted to suggest that you avoid perishable processed food that needs such dates – or am I!?)

Fruit

Supermarket fruit represents all that is good and bad in food. On the positive side is the availability of an enormous range of products from all over the world, attractively packaged and

presented. Each item is perfect and it is relatively cheap: a veritable gustatory Aladdin's cave.

On the negative side, is the impact of the measures needed to
provide such perfection. Take, for example, a box of six red
apples, each of the same size, the skin shiny, uniform and
immaculate. These apples and trees could have received as many
as 30 sprays of chemicals during growth with functions as diverse
as to pollinate, to control fungi, insects, to alter maturation and
size, and to promote growth. The ground under the trees will
also receive treatment and, at the time of picking, the labour force
will receive peanuts as far as their wages are concerned. It is no
wonder that so much fruit is imported from the Third World and
public concern over chemical sprays (remember Alar) does not
abate.

Yet people go on buying the products. It is simply not possible
to have uniformly perfect cheap fruit without sprays. You can get
organic produce from some outlets; the fruits won't be cheap,
and their appearance may be a bit ragged. It is even doubtful
whether you can taste any difference.

But there are differences among the ways that varieties of fruit
are grown. Apples attract most spraying, citrus fruit much less
and the removal of thick orange and lemon peel does much to
allay any worries. Pineapples, bananas, plums and soft fruit would
also be sprayed less than apples. Are supermarket apples really
worth the effort of eating them?

Tomatoes continue to concern many people. The genetic
engineers, use of the hydroponics method (growth without soil)
and the plant breeders have managed to develop large red-pink
tomatoes that have few seeds, much flesh, and travel well. Unfortunately they have little flavour and their function is to decorate
otherwise drab food. Modern hybrid tomatoes are hardly worth
eating. The skins are often tough.

In one open air market in Michigan, USA I was pleased to see
tomatoes described as vine-ripened, indicative that most supermarket produce is picked green-yellow and ripens during distribution. That same market in August 1992 also had a wonderful
selection of locally grown fruit and vegetables, in addition to some
imports. The maize still had its sheath and withered brown flower,

the smell of melons greeted you yards away, the blueberries, peaches, nectarines, friar plums were all being avidly bought by the customers. So too were the apples, despite their slight 'defects'.

Meat

Over the last decade, specialist butchers have progressively lost their business to the supermarkets. The consumer no doubt finds it more convenient to purchase all the food in one shop.

There are many misconceptions concerning meat. The first is what is understood by fresh. Meat eaten soon after the animal has been slaughtered is often so tough that it is inedible. The meat is therefore stored at around 10 °C for about 10 days, during which time it becomes tender, through the action of enzymes naturally present in the meat and the products of harmless bacteria growing on the outside. Secondly, controls over conditions of slaughter are now, rightly, quite emphatic. It is important that an animal is not frightened just before slaughter, for two reasons; in addition to animal welfare considerations, the meat of a frightened animal can be tough.

Shops near farms may give the impression that they are selling local meat, which may be misleading. Almost all meat belongs to a massive national and international trade, and it is impossible to identify the source of any cut with certainty. Before distribution to retailers, boxes of meat can exchange hands between dealers many times.

This does mean that the quality of supermarket meat may be just as high (or low) as that from specialist butchers. But where all raw meat on offer is ready wrapped, the choice afforded the customer may be strictly limited. The specialist butcher has the opportunity of providing the exact cuts, pieces, bones, etc. that may be wanted, including the dog's favourite. It is gratifying to see that some supermarkets are actually converting sectors within the shop to an equivalent of a specialist butcher's shop.

The meat that should give most people the greatest headache

is mince. We seem to expect mince to have the leanness and quality of finest sirloin, but to be half the price! In general, you get what you pay for, and the cheaper the mince, the greater the amount of non-meat in it, particularly fat. It is gratifying to see that some supermarkets, notably in the USA, do spell out exactly the source of the components in the mince.

One function the specialist butcher would often fulfil for his customers – certainly his regulars – is to permit them to select the cut and then mince it as required. This facility is likely to continue to decline. Will I upset you if I suggest that you mince it yourself? Tough. I have. If you do, you must clean and dry out the mincer very thoroughly after use, as all sorts of nasties might grow in the debris.

Bread

Having read Chapter 6, you will appreciate my feeling of horror over so much supermarket bread, and in my opinion the USA products are often inferior to those of the UK. No wonder bread is a commodity in grave danger of becoming extinct. It is not just the availability of the wrapped moist sliced white loaf: it is its total domination of the market.

Let us be quite clear. Real bread, like real ale, is brown. OK, there is incorporation of bits of wheat germ, or bran or rye flour, into products that produce an image of interest or desirability. But these are essentially token gestures. Particularly in the USA, what I believe is ideal bread – the unsliced, wholemeal, fresh (that is, locally baked) loaf or the white French stick – requires considerable ingenuity to obtain.

Surely we can slice our bread and produce what we want and enjoy. May I explain to you how to slice a loaf? First place it on a dry board, place a sharp knife on its top and with a downward sawing movement, lower the knife as far as the board. Yes, I know, you have more important things to do such as driving and watching television commercials, and what a dreadful mess all those crumbs make; but surely it is worth the effort? Real bread is a pleasure, I assure you, and in the chapter on enjoying your food (17) it forms a prominent position. Instead of real bread, you will find all manner of rolls, scones, crumpets and so on.

French toast (I have seen this called Texas toast) is popular at breakfast, and unless I misunderstand this subject completely, the purpose of French toast was to use up otherwise unwanted stale bread; and true French bread because of its airy texture and low water content tends to become stale rapidly, certainly within a day. Soaking it in egg and frying or griddling or whatever you want to call it converts an undesirable product into one that is quite enjoyable.

Old recipe books abound with uses for stale bread – bread-and-butter pudding and summer pudding are two examples. The problem is that if your supermarket trips are at weekly intervals you purchase wet bread that keeps well – it is never of high quality, at least not to me.

An allegory

Once upon a time there used to be a country that was Great because it stole produce from the other countries it owned. But wars occurred and those other countries rebelled and refused to supply the Great country any more. These countries were now independent.

The Great country could not then produce enough food for itself, so it dug up its trees and hedges. But the people had become lazy and when their town houses decayed, they did not repair them, and built new ones on farmland. So there was still not enough land. So they put more animals into fields. Even then there was insufficient food, so they put the animals into wooden sheds. They made more crops by spraying chemicals onto the land. The chemicals then poisoned the drinking water and the sea.

The people had become so lazy that they could not walk to their local shops to buy food. Instead they drove to more giant sheds in the country for all types of wonderful food. These were called Supermarkets and their owners became powerful and the local shops closed. The owners made lots of money and the people were happy because the food was cheap. But the Supermarkets got greedier and greedier and ordered the government to tell the people that greed was goodness. So the more money the Supermarket owners made, the less did the farmers. Eventually, the farmers were forced to feed their animals with pieces of dead animals to save money.

The food went bad, but the Supermarkets told the government to tell the people it was fine. The bad food was wrapped in more and more packets and boxes. The people were reassured by the packaging and went on eating bad food. All the used packets and boxes had to be buried in the farming land, which meant that more animals had to go into sheds.

Then the food started to make the people sick. The government said it was the peoples' own fault, so they went on eating bad food because they did not know how to produce food for themselves Then many people went mad and died. Then there were far fewer people who needed to be fed.

So those people left, who were known as vegetarians, burnt down the Supermarkets, let the animals out of the sheds, fed them proper feed and planted more trees and hedges.

After some years, more people were born, and there wasn't enough food again. So once more they cut down the trees, built more Supermarkets and returned the animals to the sheds. Then disease struck again.

This went on and on until there was no farm land left at all and nobody lived there.

12
Water, brown fizz and alcoholic drinks

Water

'You can't drink tap water' is a comment increasingly heard in the UK and the USA. There seems to be deep suspicion over tap water. Strangely, in developed Western countries which have made considerable effort to produce safe tap water, much of the population has stopped drinking it, whilst in Third World countries many people have no option but to drink what on occasion can only be called dilute sewage.

Seemingly, in any restaurant, hotel or management meeting, these bottles of water are mandatory. The epidemic is now infecting supermarkets. In one great Yuppie food shop in the UK, I recently counted 18 different varieties. There is less bottled water drunk per person in the USA, where cans of brown fluid are the order of the day.

Much of the UK bottled water comes from French springs, although some of it comes from British 'spas'. The marketing men have been getting to work on bottled water. First they made it gassy, then they transported it further and further, from cold mountainous climates. It seems now that the further the water has travelled, the more fashionable it is! Or it has been improved by the addition of just a hint of artificial sweetener and brown colour such as from old coffee.

Turning to the other giant plastic bottles of drinks littering the

shelves of UK supermarkets, another trend has been developed.
The marketing men have been taking things out of the product.
Take the cola type drinks. These are having the sugar and caf-
feine removed. The favourite tipple of my teenage daughter,
Miranda, is caffeine-free diet Coke. I know this because she
thinks my car is a mobile trashbin.

So you see what we are moving towards. Everyone will soon
be drinking a pale brown, slightly sparkling fluid! This might be
identical in appearance to that produced by some of our water
companies whose water is obtained from hilly areas; these should
be able to conquer the world with drinking water. Imagine the
advertising slogans: 'completely natural, no artificial additives,
just the colour and flavour of pure Pennine peat'. You, in the
USA, have been warned. It is on its way.

Now, I have nothing particular against bottled water; it is not
that much more risky than tap water, as far as contamination by
bacteria or benzene (this got into a famous French brand) are
concerned, the price is not more than that of lager, the plastic
bottles are not less biodegradable than any other plastic bottles,
the weight of carrying them around must surely be good exercise,
and they must provide good employment for thousands of people,
and help use up some of the world's excess natural resources to
provide energy needed for their distribution. But I'm easily
pleased. So easily, in fact, that I drink tap water.

Now, seriously, isn't this craze for bottled water total lunacy?
There are even succesful brands of bottled water that are simply
inner city tap water squirted into bottles. Let us ask, first, whether
tap water in developed countries is really safe? There are certainly
more tests and controls than with food. It has been appreciated
for many years that one of the dangers of drinking water is contamination from sewage, not just from human beings but from
excreta of wild and domestic animals, and also birds. Diseases
such as typhoid or cholera are spread as a result of sewage getting
into the water.

Regular tests are made by all water companies to ensure this
does not happen. What they do is look for the bacterium *E. coli*
in water. This bacterium is normally found in the intestine of
healthy humans and other animals so if sewage does get into the

water, then the finding of these bacteria is good evidence for contamination.

Very stringent controls over the numbers of these bacteria in drinking water are adhered to, and if these are too high, then either customers are advised to boil their water, or else the supply is discontinued. In practice, typhoid and cholera are now exceedingly rarely 'caught' in the developed world. Most instances of these infections are in people returning from countries where water purity is not up to our standards.

That is the good news. Now the bad news.

Since AIDS was recognised more than 10 years ago, it has been found that one of the common ways that the illness starts is with the patient experiencing prolonged and difficult-to-treat diarrhoea. The cause of this diarrhoea is a small parasite resembling an amoeba. It is bigger than a bacterium and has the ability to convert itself into a very durable form called a cyst. This organism is called cryptosporidium. It may also cause diarrhoea in otherwise healthy children and some adults. It is found naturally in water, and is thought occasionally to get into drinking water as a result of defects in the maintenance of reservoirs and water filtration plants.

However, disease in non-AIDS patients is rare, and although there is no treatment, it clears up on its own. Cryptosporidium is the Achilles heel of tap water!

But is bottled water any safer or better for you? Those types which are identical to tap water certainly can't be. And the general risk of bacterial contamination of still (non-aerated) bottled water is greater than with tap water. Dissolved gases and minerals in the water do enable some bacteria to grow, and if human bacteria from the bottler's fingers are transferred into the bottle, then further dangers must be present.

However, sparkling waters tend to be too acid for bacterial growth. The acidity occurs because of the chemical reactions following the dissolving of carbon dioxide used to induce the fizz. This is the sequence:

$$CO_2 + H_2O \rightarrow H_2CO_3 \rightarrow HCO_3^- + H^+$$

carbon dioxide   water   carbonic acid   bicarbonate   hydrogen ion

It is the hydrogen ion that generates the acidity. There
have been some concerns over the chemicals in spring water,
including too much uranium in one brand. There is also in most
countries inadequate legislation to monitor and control bottled
water. Certainly there are fewer checks than required on tap
water.

If you want advice on which type of bottled water to drink,
then sparkling is preferable to still, but the cost, both to you and
to the environment, is colossal, particularly with disposable plastic
containers.

Will not future historians point to the insanity of our society,
typified by our obsessive, neurotic and wasteful appetite for
bottled water, having invested so much in the provision of tap
water? Yes, I know you don't like the taste of tap water. Nor do
I if there is chlorine in it. There is absolutely no need to add
chlorine to tap water; the water should be collected, stored and
distributed with care, and not with chlorine.

Fermented drinks

Beers, lagers and ciders are among the most profitable items of
the food industry. These drinks are technically food, the nutrients
being almost entirely carbohydrate! Alcohol itself is almost a
carbohydrate, being composed of carbon, hydrogen and oxygen.
In the body it can be used as energy or converted to, and stored
as, fat.

The reason for the usual safety of fermented drinks is that
after the addition of yeast, the sugars and starches present are
changed into both alcohol and acid. Acidity is measured on a
logarithmic scale, known as pH. A neutral pH, meaning that the
fluid is neither acid nor alkaline, is given a figure of 7.0. When
acidity rises, the pH drops so that each drop by a whole unit
represents a 10-fold increase in acidity. Therefore a pH of 4.0
is 10 times more acid than a pH of 5.0 Most bacteria, but not
yeasts, stop growing at a pH of 4.5 or less. Nearly all beers, lagers

and ciders have a pH between 3.7 and 4.2, and so prevent most bacteria from growing, certainly those that cause food poisoning. However, bacteria found in the air and in the environment can grow under these acidic conditions and this explains why opened bottles or cans go 'off', and why cleaning the pipes between barrel and pump is important. The alcohol of these drinks is too weak to act as a preservative. If mistakes do occur it is likely to be with the early stages of fermentation.

Home-made beer sometimes gets into difficulties, if it is made under dirty conditions that enable bacteria to overgrow the yeast with the generation of products such as acetaldehyde. The brewing yeast has been adapted to grow at room temperatures. At the end of fermentation, any remaining yeast drops to the bottom of the vat or barrel and the drink clears, unless it is disturbed. However, there will probably be some yeasts invisibly present even in the clear drink. These pose no risks to health because brewing yeast cannot cause illness in us. For a start, we are too warm. Human infection from yeasts is due to strains that grow well at body temperature (37 °C) but not at room temperatures. The word yeast covers an enormous variety of tiny organisms, the great majority of which have been used by us for our benefit. These can also be used to make wine; but some may cause an unpleasant taste, if they are introduced into the fermenting grapes.

Low-alcohol beers and lagers can be made in one of two ways. The product may be prepared by fermentation in the usual way, and then the alcohol removed, for example by boiling. Then the acidity of the product should be high, so it should be safe and not need preservative. If the product has not been fermented sufficiently to generate adequate acid, then problems can result unless preservatives are added. Sulphites may be used. In the brown cola type of drink, a favourite chemical preservative is benzoic acid.

The amount of alcohol in beers, lagers and ciders varies from about 2–5% by volume, and that of wine 9–11%. Different yeasts have been adapted to survive the higher alcohol concentration.

Despite complaints about dirty glasses and premises, I cannot identify a single instance of poisoning from contamination of commercially produced beer. Indeed, if you put salmonella bacteria into beer, they die. Put another way, beer is an antiseptic! Products from hops also contribute to its antibacterial properties, so the brews which are the most potent antiseptics are real bitter ale and strong cider! For alcohol itself to prevent bacteria from growing, concentrations over 15% are needed as in sherry, port or spirits.

Cloudy beer is usually due to one of two problems. Either the yeast in the bottom of the barrel has become disturbed by rolling, or secondary bacterial growth has occurred by neglecting to flush fluid through the pipes between cellar and pump.

Spirits, with an alcohol content of 35–40%, therefore keep almost indefinitely, even after opening. It is interesting that the optimum proportion of alcohol in water as a disinfectant is 70%.

Effects of alcohol

The effects of alcohol have been well described and it is not the intention here to go over well-trodden ground. There are some points, as follows with which everyone may not be familiar.

1. The immediate effects on the persons are due mainly to the rate of absorption of alcohol into the blood. Impairment of brain function is due to the difference in the alcohol concentration between the blood level and that inside the brain cells. Some hours after stopping drinking, the brain and blood levels will have equilibrated to be nearly the same and unless the levels are very high the effects from the alcohol will be wearing off. This does mean that the relationship between the amount of alcohol drunk and the severity of its effects on you is often not close.

The law usually relates to measured levels rather than to the effects on the individual. This is because it is possible to measure the amounts of alcohol in blood, urine or breath rapidly and accurately, but it is often not possible to measure the actual effects on the person at all easily. You can demonstrate a deterioration in behaviour only if you have documentation of what the behaviour is like under 'normal' conditions.

The effects on the brain vary from individual to individual. A common effect is to dampen down some of the function of the brain involved with negative controls. This is seen as removal of inhibitions and is apparently the main *raison d'être* for social drinking. What a pity we have so organised our society that such negative behaviour is usually dominant!

There are claims that mixing drinks is more liable to produce adverse effects than drinking a single alcoholic beverage. There is little scientific evidence for this; perhaps the key lies with this concept: the greater the variety of drinks consumed over a finite period, the greater the total amount of alcohol drunk?

There are numerous claims for hangover treatments. The only possible relevant action might be to replace the body's water lost as a result of the alcohol's action on the kidney. Typically, for every two pints of strong beer drunk, two and a half pints of urine are produced. In medical terms, alcohol is known as a diuretic and makes the kidney produce more urine than it would do for the same volume of non-alcoholic drink.

Labelling does enable consumers to know the amount of alcohol in the various drinks. As a guide-line, if there are no labels, low-alcohol long drinks are 1% alcohol; beer, cider and lager, 3%; wine, 10%; sherry and port, 15%; spirits, 40%. Much has been written about 'safe' levels. Let us look at these first as regards driving.

After some time has elapsed, even up to 24 hours after drinking a substantial amount of alcohol, the amount present in the body can still be at illegal concentrations (from the driving point of view) despite the ability of the person to appear to function quite normally. This accounts for many of the infringements of the law of the 'morning after the night before' type! It is not possible to give hard and fast figures, and on no account rely on these because of individual variables. But typically, the blood alcohol of an average 40-year-old male after drinking 3 pints of beer will peak at 80 mg/100 ml. After 6 pints, the figure is 160 mg and it will take 8 hours for the level to drop to 80 mg, based on the premise that the body can eliminate 10 mg equivalent alcohol every hour.

2. Even with low-calorie or 'lite' beers or lagers, the alcohol is calorific and has either to be burned off as energy or, if it is not needed, is converted to fat. Most long alcohol drinks generate 200–300 calories per pint.

The question of the safe amount of alcohol that can be taken in the long term is not at all easy. There is enormous variation in the ability of people to tolerate alcohol. But we do know some of the factors involved. Women as a group tolerate alcohol less well than men. (Sexist, but true!) It is difficult to know why, because there is no real scientific evidence. It might possibly be partly related to the generally lower weight of women relative to men.

Then we can identify three phases of toleration in a regularly drinking person – say, 6 pints of beer daily. The first phase is at the beginning, shown by the low threshold to experiencing immediate toxic effects, such as staggering, slurred speech, amnesia, and aggressive behaviour. This is followed by gradual development of tolerance, where the amount of alcohol absorbed into the body is maintained, but the effects of it are reduced. Then the third phase, often associated with liver disease, is a return to the low threshold.

Alcohol is a general tissue poison and there is good medical evidence that long-term consumption of moderate–high amounts of alcohol can damage the liver, heart and brain. Sometimes the effect on the brain and nerves is seen as curious patches of numbness over the skin, or straight physical dependence: that is, the person must continue to drink alcohol.

When thinking about safe amounts, much has been written suggesting that the regular intake of moderate amounts of alcohol is more dangerous than the occasional binge. I am not aware of much evidence to support this claim, and of course the binge can result in death from vomiting, inhaling the vomit and asphyxia.

Certainly, there are many people who regularly take a large amount of alcohol without coming to any obvious harm. A very important early warning sign occurs when a person's lifestyle is altered in order to ensure that his or her alcohol intake is maintained.

In addition to the various diseases specifically associated with a high intake of alcohol, there is a general excess mortality, for reasons not fully understood. So the heavy drinker is more likely to suffer from cancer at, say, age 55 than other people.

Bearing in mind all these doubts, it is only very tentatively suggested that safe daily limits can be identified; 1–2 pints of beer, lager, or cider proposed for women and 2–3 pints of these drinks for men should not be set in tablets of stone.

Pregnant women

Being pregnant immediately puts you into a new, high-risk category! Research performed largely by men, has shown some not very convincing evidence that alcohol consumption is likely to have a deleterious effect on the baby. I say, not very convincing, because the research relates to fairly high consumption. Nevertheless, the Surgeon General in the USA has issued warnings, as shown by wrappings around bottles of fairly weak beer, as follows: 'Women should not drink alcoholic beverages because of the risk of birth defects'.

This would seem to be a somewhat draconian warning from the evidence available, and makes no comment of the amount of alcohol drunk, nor the stage of pregnancy during which such consumption occurs.

There is no such warning in the UK. Now, I am not in favour of alcohol in pregnancy; indeed, no doctor can recommend almost any drug or food to a pregnant woman. But I do feel there is a need for more research with proper controls and to look at other aspects of the woman's culture in addition to alcohol.

With smoking, the final proof of harm was obtained by showing that stopping smoking reversed some of the risks. This, to my knowledge, has not been established with alcohol. The jury is still seeking more evidence! It is waiting for the summing up by the defence counsel!

Beneficial effects of alcohol

Research has shown that moderate consumption of alcohol – of the sort of amounts just mentioned – can have positive benefits by reducing a person's liability to heart attacks. It remains to be seen whether this apparent beneficial effect of alcohol on preventing heart attacks is offset by any deleterious trends. After all, we all have to die of something, and if we are less likely to suffer from heart attacks, are we not more likely to suffer from diseases? To my knowledge the research looking at these issues has only just begun, and the key criteria are not just the disease causing death, but the age of death and the quality of life preceding the final illness.

If alcohol is taken in moderate amounts, in the belief that it might be beneficial in reducing the likelihood of heart attacks, it must be remembered that alcohol is very calorific.

Why are low-alcohol and soft drinks so expensive?

Since alcohol is fairly expensive to produce and attracts a variety of taxes (alcohol is an irresistible target for finance ministers) why are not low-alcohol or soft drinks much cheaper than, for example, beer?

The first reason is that the price is based on what people are prepared to pay. No matter that the cost of the contents of a can of brown fizzy fluid is one cent or one penny, the selling price is one dollar or one pound, if you are prepared to pay that amount. You are! So that is the price. If you don't like the price, don't buy. Drink tap water.

But there are other reasons. First, other taxes associated with production and selling may not be influenced by the alcohol content. Then there is the cost of the package, the advertising, the hype, the television commercials.

Commercials do work, but for them to succeed really well, the product must be widely available, and large numbers of the viewing public have to be influenced. This argues very much for large-scale production, often in one or a few factories, and a national network of distribution. It is no wonder that so many of our convenience drinks are produced by a handful of manufacturers who have the ability to set any price they believe they can get away with.

13
Take-aways

The burger

What would the Romans have made of the beefburger? Would they have eaten it themselves or fed it to their animals? Historians of the next century may well look back on the late 20th century as the period when the world suffered the burger epidemic and its disastrous results. How did it all start? The history is somewhat blurred, but the account by Jeremy Rifkin in his excellent book, *Beyond Beef*, is convincing.

The development of beef cattle occurred in the flat plains of what is now Eastern Europe and Russia. There, the medieval Tartars tended to add onion extracts and seasoning to raw beef. This has given the name of 'Steak Tartar' to types of beef eaten raw (horror!). German traders brought the recipe back to Hamburg, where the steak mixture was cooked. From Germany, the recipe for the cooked processed beef went to the USA, apparently being first eaten between slices of bread at a county fair in Ohio in 1892. The Hamburger era was born and its popularity can be understood.

It could be eaten with the fingers, so tables, chairs and utensils could be dispensed with; it was readily and rapidly available and the filling, being minced beef, was popular among a population with poor teeth. There was a further aspect, which is probably more important today than previously: the processing of the meat removed any resemblance to the large slabs or joints of meat from the slaughtered animal. There has always been a need to dissociate the food on the plate from thoughts of the live animal, its slaughtering and butchering. The new shape and texture of the hamburger bore no similarity to any of the animal's organs – or to anything else, for that matter.

Then there is the name. The German city of Hamburg does not immediately have association with any live animal, unless it's the pig through the prefix 'Ham'. But over the years the 'Ham' has tended to have been dropped, so the genre is now just 'burger'. Or was. Recently other ingredients have been added to create the cheeseburger, the eggburger, the fishburger, brunchburger and so on. There has therefore been further remoteness between the name of the food and the source of the meat. Indeed, 'burger' has come to mean any type of chopped sticky material in a flat patty shape that is cooked and held between two pieces of bread material. Beanburgers are good examples of this.

One of the key requirements for the original beefburger was for the pieces of minced beef to adhere together so that, in placing the raw patty on the heating plate and then incorporating it after cooking into the bread bun, the patty does not disintegrate. The German origin suggests that considerable amounts of fat are required. (German sausages and processed meats have tradition- ally been high in fat.) This means that the typical beefburger will consist of around 20% animal fat.

Much of the expansion of the burger trade in the USA between 1920 and 1940 was related to the developing network of roads and motor-car ownership. In the last 30 years, one particular company became dominant – McDonald's – and as far as quality is concerned, their products represent burgers at their best, and the company has given great attention to detail. The size of the beef patty and that of the bun, and the amount of onion, are carefully portion controlled. The bread contains a little more sugar to speed up browning during toasting, and the minced beef would, for example, contain no heart or lungs of the animal.

One of the factors crucial to the development of the beefburger was the ability of the meat patty to survive deep-freezing. As with most processed foods high in fat, the texture is not substantially damaged by freezing, as long as it is eaten immediately after cooking from frozen. This, of course, is virtually inevitable with the burger: its entire *raison d'être* is its instant consumption. Try keeping a cooked burger and its deterioration is rapid indeed.

In theory, the thorough cooking of the frozen patty should remove dangerous bacteria. Unfortunately this does not always happen, particularly with large burgers, and food poisoning due to the bacterium *E. coli* 0157 (see Chapter 9) has resulted from burgers in several American states and in the UK. There are also doubts over the nutrition of the burger, concerning both what is in it and what is not in it; and the worry that the seduction of the burger, being so available, encourages overeating. It is impossible to assess the nutrients of a product in isolation from the rest of the diet, but a diet based mainly on beefburgers would run the risk of shortage of some vitamins, such as C, shortage of polyunsaturated fats and fibre, and excess saturated fat and carbohydrate.

Whilst the McDonald's burger represents the gold standard for the beefburger, developments more recently, particularly in the UK, have resulted in some very poor quality and even danger-ous food.

Degeneration of the burger

The first development was not to bother to toast the bun, which was now limp. True, some were hard-baked and sprinkled with all manner of seeds, but the untoasted bread in contact with the fatty patty absorbed the fat rapidly. The move to the soggy burger had begun.

Then the quality of the meat deteriorated. 'Beef' refers to any edible part of the cattle, and does not necessarily mean lean meat. Burgers, like sausages and meat pies, have become wonderfully useful vehicles in which to dispose of otherwise unsaleable parts of the animal.

Cooking is now often uncontrolled and haphazard. Because of the risk of *E. coli* 0157 in inadequately cooked burgers, the UK Department of Health issued a warning to the public in 1990, asking them to make sure that burgers were well cooked.

Then more and more additional items were put into the
bun: cheese, egg, bacon, limp lettuce, tomato ketchup, brown
sauce, almost anything. The burger patty became smaller and
fattier.

Then the double-sized variety appeared – called King,
Jumbo, etc. – proof indeed of the shrinking size of the 'standard'
variety.

Worse was to follow. The appearance of the microwave has
been decisive in altering catering methods. The microwave energy
passes straight through cardboard or polystyrene boxes, so the
scene was set for the reheated burger to be found on trains, street
corners, restaurants, garages, canteens: everywhere.

The burger patty is cooked in advance, chilled, inserted between two pieces of untoasted white bun, put into a box decorated
with pictures of wonderful burgers and stored, transported,
stored, transported, stored and so on, hopefully under carefully
refrigerated conditions. The punter, on seeing the garish picture
on the box, finds the idea of the burger irresistible. He pays
through his nose and, after a minute's mauling in the microwave,
he has his most marvellous snack. The hot, damp bread, remi-
niscent in texture of cotton wool, contrasts excitingly with the
gritty, leathery, reheated patty. It probably won't contain any
salad, which would only get in the way of his enjoyment.

But it may contain a slab of orange-coloured sticky material,
stated to be cheese. The typical UK punter has heard all about
BSE and that it might be in beef but may be unconcerned because
he or she may not actually appreciate that a cheeseburger contains
any beef.

What matters is the convenience. It is now a requirement for
modern youth that, on the few occasions he is required to walk,
he has also to eat. Munching between outlets of microwaved
burgers provides the means. Moreover, he has the pure unmiti-
gated pleasure of dropping the polystyrene box onto the ground,
giving it a kick as if it were a football and seeing it come to rest
in the gutter. He has, like a dog lifting his leg to a tree, left his
mark on the environment.

Other microwavable delights

The fast food floodgates are now well and truly open. Almost anything can be cooked in a centralised factory, cooled down, bunned and carted around the country to all the microwaves. Already convoys of lorries are on the move. Bacon buns, eggs, pork pies and spaghetti have all been tried, but the real delicacy is the chip butty, so beloved by the sophisticated taste buds of Yorkshire men (and women). This glorious product is exactly what you think it is. Chips (not crisps, but a species of french fry) are fried in beef dripping at fairly low temperatures to ensure that it can be used many times and the chips or fries are satisfactorily soggy. Then, hey presto, the bun (white bread, of course) is cut open, the fries rammed in and, after storage for about a week, the microwaving ensures perfect disintegration of the fries.

This is so popular in the North of England that it is even served on plates in pubs and clubs. And for the truly discerning foodie, it can be accompanied by mushy peas – which, for the uninitiated, represent the pinnacle of haute cuisine. These are farmed peas which have become too large to be frozen and are already fading in colour. During the canning process, they are heated for sufficiently long to ensure that much of the vitamin content has been destroyed, they have changed their colour to khaki and their texture to something between lentil soup and lumpy porridge. Stale pork pies are also fashionable as an accompaniment.

Instant chicken

Much intensively reared broiler meat goes into instant chicken, recognised by its pale colour and presence of small, brittle bones in drumsticks or wings. Some of the poorer quality residual meat is minced and fashioned into patty shapes to make into chicken 'burgers'. Other products are shaped into lumps and then coated.

There are some practical problems for the provision of cheap instant chicken. As a result of the slaughtering and preparation methods, it has to be assumed that there is a high likelihood of salmonella and/or campylobacter bacteria being found on the outside and inside of the carcasses. Small pieces can be covered with coating and stored frozen. The subsequent cooking in hot oil or fat should enable thorough heat penetration of the item (and therefore make it safe) before the coating burns.

With larger items – drumsticks, wings or halves – the problems develop if these are cooked, frozen, and then cooked again. The coating is likely to burn before thorough cooking has been completed. The following options attempt to solve the problem but are not entirely successful.

The chicken might be cooked – say, by steaming – before

it is coated and frozen. After frying it will probably be safe, but in an advanced state of disintegration from its double cooking.

The raw, coated frozen products might be allowed to thaw out – or, indeed, be microwaved to accelerate the process – before cooking. This will only slightly aid heat penetration before burning of the coating.

Or the uncoated frozen chicken pieces can be thawed before coating and cooking. But this requires a good deal of local activity which is much more costly than centralised production. Similarly, the coating of fresh, unmanipulated chicken portions at the local eating house is also more expensive.

Then, finally, the chicken portions might be treated as follows: raw pieces are cooked by steam, then cooled, dried and coated, and stored chilled (hopefully at a temperature below 5 °C). They are then briefly cooked in hot fat. The safety of this system depends on the certainty that storage temperatures are cold enough (ideally as close as possible to 0 °C)and are not too long – ideally no more than 3 days.

Other options do exist, and it is obvious that the safe provision of high quality instant chicken requires stringent controls and an understanding of the whole process. Overall, fast chicken must pose substantial risks, because of the complexity of the range of the procedures used, and the risk of human or equipment error. Like the beefburger, the cheap copiers of the real thing are most likely to run into trouble.

Fish and chips

Surely fish and chips, the great favourite of the British must be beyond criticism? I would give them fairly high marks, say 7 out of 10 overall, but I would comment on their report, like that on our crops, 'could do better'.

Fish and chips (that is, a variant of French fries) are generally cooked to order, and should be just about as safe (microbiologically) as any food. True, there has been one report of some fish being contaminated with a disease similar to typhoid from the fingers of a server. But so could any food; it is not the fish's fault, it is the human's!

Most of the fish fried is certainly unprocessed, but the claims for freshness are a little more contentious. The description of a food item as 'fresh' means different things to different people. I suppose to most of you it means that the product is in the condition in which it was first picked, harvested or caught. It is not frozen, dried, canned or reconstituted.

But let us look at the reality of the fish trade. Most fried fish is of the deep-sea white variety, particularly cod, plaice or haddock. The boats may well be out of port for a week or longer, so much of the catch is frozen on the boat and distributed later whilst still in a frozen state, to be thawed either before delivery to the frying premises or actually on site. Fresh usually means frozen and thawed!

Fish does survive freezing and thawing relatively well. I know of one smoked salmon business in the South of England that buys frozen farmed salmon from Ireland, then thaws, smokes and slices it, prior to packing and freezing in preparation for export to the USA. There it is thawed prior to sale and for all I know it can be used to fashion food items themselves to be frozen and ultimately thawed for the third time.

There is another problem when the term 'fresh' is applied to fish. In the most exact sense, the freshest fish are those that are available, cooked and eaten, immediately after catching. There is a cult following for this. But is the taste at its best immediately after catching? I am afraid it is not: the taste and smell are often over-bland and nondescript. The most positive flavour of fish occurs after 2–3 days' storage at 10 °C, but much longer than this results in a bad smell, often ammoniacal in nature, and other signs of decomposition such as dullness of the skin and opacities in the eyes. The reason for these changes – that is, first the development of a positive flavour, to be followed by 'off' smells – is the growth of the so-called food spoilage bacteria, such as pseudomonas. The practical problem with a fishing boat at sea is that it is usually not possible to plan the correct interval of storage to generate these flavours. Hence the fish are usually stored frozen.

What about the oil? In most shops, the oil is vegetable oil, hopefully changed regularly to prevent chemical changes, known as denaturation. For completely inexplicable reasons, most Yorkshire fryers use beef dripping. This produces a fattier product. The fish is always covered in batter and I am not quite sure why batter is used rather than breadcrumbs; certainly batter is easier and quicker to use, but do the customers prefer batter to breadcrumbs?

The chips (fries) are very variable and surely the variety of potato dishes could be improved? Like most take-away food, chips deteriorate rapidly with time, so the specialist fish restaurants are justifiably popular. Here the clients are obliged to become familiar with plates and cutlery – what a hassle!

But as with burgers, the genuine article has become devalued. There are chains of restaurants that receive deliveries of frozen, battered fish and the customer may well be in for a rude shock, particularly if it is a large piece of fish. The batter can be golden and crisp and well cooked on the outside, but the inside of the fish could be raw and still frozen! We seem to be becoming lazier and lazier with food preparation, largely driven by the extraordinary reticence of the customer to complain!

See what I mean? Fish and chips could do better.

Sandwiches

Would the Earl of Sandwich now be proud of what he has started? Sandwiches have invaded most of the USA and the UK. Not just the English cucumber sandwich for afternoon tea in five-star hotels, but a myriad of sandwich types are increasing to epidemic proportions everywhere. Filling Stations (garages) used to sell mainly gasoline, but I am told there is often more profit from food sales. This began as chocolates and sweets, moved to crisps (chips) and cookies, and now it is sandwiches and anything capable of being ruined by the inevitable microwave, such as pies.

There appears to be some confusion as to what is meant by the term 'sandwich'. This is important, as there is strict legislation as to what can and cannot be done to sandwiches. The flattened circular items of bread that I call baps or buns are referred to in Yorkshire, UK as breadcakes. When these are sliced open, a dollop of grated cheese and coleslaw gunge squashed by hand between the top and the bottom, and placed in a cellophane bag, it becomes a sandwich. But in general sandwiches are made from sliced bread, usually purchased in that condition!

Not that I am against all sandwiches. Some are really excellent, but with such a widely available commodity often in rigid packages, we tend to forget that the bacteria in or on the food are alive and that we must therefore take care in the making, storage and eating of sandwiches.

There are numerous laws and regulations that attempt to ensure sandwich safety. The experience of the writer is that these are regularly flouted, and are not capable of being enforced in practice because of the difficulty of establishing the details of time and temperatures of storage of all the possible ingredients. So let us try and understand the underlying principles to achieve a safe sandwich, as follows.

1   The bacteria responsible either for contamination or for food poisoning will not grow to any extent in bread, traditional butter, margarine, hard cheese or pickles in these foods in their pure state. They will also find the going tough on *whole* salad items.

2   Cooked meat, fish and egg provide excellent food for bacteria.

3   When food components are chopped or intermingled, the opportunities for bacteria to begin to grow are usually increased. For example, a sandwich containing sliced (or grated) cheese with cut tomato and cucumber will allow enough water to enter the surface of the cheese to make it ideal for bacterial growth.

4   The number of bacteria put into a sandwich in the first

place will govern how long it may be kept safely. Mass-produced cooked meats remain the greatest single source of contaminating bacteria.

5 But even if in the making up of a sandwich the combination of different foods provides ideal growth conditions for any bacteria, this does not mean that they will start growing immediately, even if the sandwich is kept on the warm side. In practice they will not start into growth for, typically, between 2 and 6 hours at temperatures between body heat (37 °C), the most rapid onset and, say, 20 °C (typical room temperature), the slower start.

6 To stop bacteria growing at all, if it is to be kept for say longer than 4–6 hours before being eaten, the sandwich should be kept really cold, as near to 0 °C as possible.

In practice, then, we can suggest the best ways to make and store sandwiches.

For the seemingly shrinking number of people who make up sandwiches for themselves or their family, there is nothing wrong at all in preparing the sandwich for lunch in the morning before going to work or school and that sandwich being kept for a few hours at, say, 20 °C. Similarly, if the local sandwich shop makes up sandwiches during the morning for the lunch trade, no problems should arise, unless of course the ingredients are defective.

If the school packed lunch is prepared the evening before, the sandwich should be refrigerated overnight. One point of concern can be the cleanliness of the sandwich box: it is essential that it is cleaned out with soap and water (no need for antiseptics), rinsed and dried each evening.

I think it is well worth trying to eat sandwiches which have not been refrigerated because of their better and more positive flavour and texture. Those which are centrally mass-produced have to be refrigerated in order to survive their long, tortuous distribution and storage network, and the results are often wet and tasteless. Do you think you can distinguish a stale grated cheese and cole-slaw sandwich from a stale chopped prawns and mayonnaise one? Try it blindfold!

Many of these centrally produced sandwiches have sell-by or eat-by dates up to 3 days hence and if kept really cold (i.e. near to 0 °C) they are probably safe; but how often do cold stores break down, and how often do people, including yourself, forget to refrigerate food? Is the cold, damp, tasteless product really worth eating? Consider, too, the environmental impact. Because of the distribution procedures, most of these types of sandwiches are enclosed in rigid containers, and there are energy costs for the refrigerated lorries, stores and displays to be considered. The problem is not just that of energy but the coolants in the refrigeration units. These are still predominantly chlorofluorocarbons (CFCs) and I need not remind readers of the effect of these on the Earth's ozone layer.

The argument for locally produced, fresh sandwiches that do not require refrigeration seems compelling. Are not these more enjoyable, safer, environmentally friendly and probably cheaper? Yet common sense does not seem always to prevail. Organisations seem to find it convenient to enter into comprehensive contracts with individual suppliers.

All sandwiches available on British Rail in the UK have been assembled in a single factory in Staffordshire. The ingredients going into them may come from as far afield as Scotland or Germany. Then, after assembly, a fleet of vans takes them overnight to the main-line terminals all over mainland UK and from these to the trains and local stations for the hungry traveller next day. They are not cheap.

Why can't sandwiches be prepared at each station at suitable timings for the required trains? I regret to have to say this again: stale sandwiches are more prevalent in the UK than in the USA.

The Oriental take-away

Seen from the view of the customer, the near instantly available food is usually ushered in to the reception area from an out-of-sight type of kitchen. He or she does not usually question how it is produced. The key factors are that the food is provided fast and hot.

But seen from the view of the restaurants, this customer is but one of many (hopefully) who will come in at unpredictable times and groupings over many hours, even throughout the day. How is such a demand met? The answer is not at all easily and dangers can exist.

Take Chinese fried rice, probably the most hazardous item. This will be cooked in bulk as boiled rice, in preparation for the day's opening. This can be the morning of the day in question or the evening of the day before, and it may be kept for more than one day.

The bacterium *Bacillus cereus* is often found on rice. When the rice is dry, this bacterium will stay mainly in its spore state, meaning that it is well adapted to survive boiling and therefore be invisibly present on the boiled rice in the giant saucepan.

The first problem occurs with the time taken for this volume of rice to cool down: it may be long enough for the *B. cereus* spores to start turning into growing bacteria that may well release toxins into the rice. But let us hope that this will rarely happen and that the rice is transferred to a well controlled refrigerator.

The first punters come in and order fried rice. The vessel of boiled rice is taken out of the refrigerator, a scoop of it is put into a frying pan, perhaps some egg and shrimps are added and the rice is stirred briefly to heat it to 50–60 °C and transfered to containers. It is eaten with relish. No problem.

But a succession of people come in and there is simply no point in putting the cooked rice in the refrigerator every few minutes because each time the door is opened, the fridge warms up. In practice, the boiled rice may be left at room temperature for hours, the bacteria grow and the last customers get a nasty dose of food poisoning. One way round this problem is to batch the boiled rice, refrigerate the batches rapidly and, indeed, to have more than one refrigerator so at least one batch stays cold.

The problem here is typical of the general risks associated with the need for instantly available hot food.

It is odd that fashions in food are such that we should want it to be at almost any temperature as long as it is not at body temperature; hence the vogue for hot food and ice creams. It seems virtually impossible to purchase a drink in the USA without the glass being filled first with ice cubes!

Is there a slow death from fast food?

The practice of intermittent eating of small amounts of fast food over the course of the day has become known as 'grazing'! There are a number of misunderstandings in this area. The first issue is the belief that six small 'bites' daily might be less desirable than three medium ones or one enormous one, French fashion. There really is very little evidence that the frequency of eating has a major effect on long-term health. But the other concern is that fast food often contains too much fat, sugar or salt that are needed in order for it to be fast. We have seen that burgers have a high fat content, as do sausages and hot dogs. The popularity of chips (fries) must in part be due to the feasibility of eating them with the fingers. Eating mashed potatoes with fingers is not likely to become popular. So in several ways fast food favours certain types of diets, and as a consequence introduces risks, in at least some people. But it need not, and already there is evidence of some improvement in the quality of some types of fast food. Do we not detect more salads in sandwiches? This is probably one of the food areas where the UK is outgunning the USA.

There also seem to be available more fruit, quality fresh pastries, recently baked bread baguettes and fishy sandwiches, all of which are welcome. Perhaps one of the ideal fast foods is the pizza made and baked to order. Yes, I know it is not truly fast, taking between 5–10 minutes, and it would be undesirable to keep the car engine running for that time.

In conclusion, fast food can be of high quality, can be nutritious, even enjoyable and safe, particularly if locally prepared. Frequently, however, fast food represents everything that is undesirable, being environmentally damaging, fatty, poor quality and expensive – but hot!

14

For those of you with kitchens

This chapter is written for those of you who still cook in the kitchen, or for those people who would like to face up to the challenge but do not know how to do it safely. The tips on food safety at the end should easily become automatic, and it is not the intention to deter anyone from home cooking. Quite the opposite: it really can be fun.

Refrigerators

I recently visited a refrigerator showroom in the UK. It was bursting with new refrigerators. I asked the assistant how many of the models had thermometers installed. She didn't know, so called the second assistant. She was not sure either. While the manageress was being sought, I thought the best way of answering the question was to look inside each model.

Soon we were all in agreement. Not a single new refrigerator contained any type of thermometer. My next question of the ideal temperature stimulated a conference. Then the manageress replied, 'Different manufacturers recommended different temperatures, some say between 0 and 4 °C, others between 3 and 7 °C, and others up to 10 °C'. It occurred to me that at least they think that some temperatures are desirable and others are not.

The question, 'How do you know whether these temperatures

are reached, when there are no thermometers?' drew a blank with the first assistant. The second assistant came to the rescue, opening a door. 'You see this dial with numbers 1 to 7; each number is the temperature you want'.

There seemed to be some confusion here. I tried to explain. 'The dial setting controls the activity of the motor responsible for cooling. It does not tell you the temperature reached. That depends on the warmth of the room air, the closeness of door seals, the frequency of opening, the amount of food inside, and the position of the thermometer in the refrigerator with the coldest air sinking to the bottom.'

The glossy brochure was then introduced. There were 31 different models listed and illustrated. There was no mention of *any* temperature anywhere. Moreover, the use of many of the fridges was clearly unsatisfactory, if not downright dangerous. Few items were wrapped. There was no separation of raw and cooked products. Some of the raw unwrapped meat was right at the top. This can be dangerous on account of juices dribbling from, for example, a raw chicken onto food underneath which could then become contaminated with salmonella, campylobacter and listeria.

So how can we account for this complete shambles? Why has the recent publicity over the dangers of chilled foods had no impact on the manufacturers? Why did the new UK Food Act not require thermometers to be installed?

One answer, I suspect, is that several models cannot actually achieve a safe refrigeration temperature for cooked food. This should be as near as 0 °C as possible. A thermometer should be kept in the middle of the cabinet and checked on first opening the door each morning. Another reason could be that the makers of refrigerators like to be vague about the temperatures in order to encourage their use for all sorts of purposes.

Incidentally, a number of foods that have sometimes been stored in refrigerators are preferably kept cool at, say, between 10 and 15 °C rather than really cold. These include eggs, unopened cheeses, salads and whole raw vegetables. Prepared (particularly if chopped) salads and coleslaw should be kept cold – as near to

0 °C as possible. If you cannot achieve cool temperatures, then the refrigerator may have to be used (Table 3).

Table 3. *Safe management of the home refrigerator – recommended storage times (0–3 °C)*

Food type	Storage time
Raw food	
Whole joints (not processed), to be cooked thoroughly	3–5 days
Sausages, mince, offal	1–3 days
Bacon (no preservative)	3–5 days
Bacon (with preservative)	According to packet instructions
Prepared vegetables and chopped salads	1–5 days, according to appearance
Fish	12 hours
Cooked food	
Meat, poultry, ham, pies, stews	2 days
Vegetables	Avoid
Gravies, sauces, custards	Never
Opened cans	2 days
Dairy products	
Milk and cream (pasteurised), unopened	2–3 days
Milk and cream (pasteurised), open	2 days
Cheeses	According to packet instructions, or for as long as the appearance is acceptable

Other aspects of safe refrigerator management are as follows:

Check door seals weekly to make sure they are not defective.

Wrap individual items in transparent material, or label opaque containers.

Place raw items on the bottom, cooked or ready-to-eat food at the top.

Clean out thoroughly, weekly.

Do not crowd items too closely together.

CFCs and the ozone layer

There seems to be some confusion over the two quite separate causes of the global environmental crisis. Global warming is due to excess gases, such as carbon dioxide and methane (incidentally, the fitting of catalytic converters to cars will do nothing to help global warming because they actually increase the amount of carbon dioxide synthesised from the poisonous carbon monoxide).

The basis for refrigeration is to allow a fluid to condense, a process using up heat and so producing cooling. The cycle is made continuous through evaporating the condensed liquid by heat. This does incidentally mean that operating refrigerators produce a net gain of heat, but the large room space is warmed up only slightly, although the inside of the proportionally small refrigerator is cooled down markedly. The coolant used for this purpose within the closed system adjacent to the refrigerator chamber has to possess critical properties of being capable of alternating being in gaseous and liquid phases under the temperatures and pressures that are achievable in the home. Very few substances are suitable. Chlorofluorocarbons (CFCs) have been used almost exclusively because they were found to be the only ideal substance. Ammonia is less efficient: for the same amount of cooling, much more energy is needed. Other substances are being tried, but are still in the experimental stage.

By their very nature, any coolant that can evaporate can escape, with a probability that it will end up in the atmosphere unless extraordinary precautions are taken to prevent it. The damage to the ozone layer is real and is due to escaping CFCs. Fortunately, were CFCs to be reduced soon, the ozone layer should recover over the years because it is being constantly regenerated from oxygen interacting with various types of irradiation bombarding our planet.

We use refrigeration (and the deep freeze – see Chapter 6) mainly to preserve moist food, and although the refrigerator seems essential to living, it was not always so, being an invention of this century. In Victorian England, wet food was not stored for any length of time. For the average household, the cool of the cellar or the pantry was the best that could be provided, and houses were built with north-facing kitchens to lower the room temperature. The very rich brought in loads of ice to chill their wine.

Many of our problems, particularly environmental, have been due to our determination to store moist food. We have already seen that many other types of food preservative are satisfactory, and that if we cannot reduce our emissions of CFCs, we may have to redesign the household refrigerator and deep-freeze. (If I could choose just one of these, I would go for the deep-freeze!)

Microwave ovens

Microwaves typify everything good and bad about our attitude to food, indeed to our society. I am not against microwave ovens *per se*; their value depends on how they are used. What concerns me is this: much of our preparation of food is being geared to the availability of microwave ovens. I know of an entire student residence where the only cooking facility in each kitchen is a microwave oven.

More microwave ovens have been sold in the UK per head of population than any other country, including the USA, attesting to the new British character featuring impatience, laziness and greed.

Microwave ovens work by emitting short wavelength electro-magnetic irradiation from a magnetron. These are in effect the same as military radar, and are directed at the expected position of the food. Many miss their target, and bounce around the inside of the chamber until they strike the surface of the food. The microwave energy is dissipated when it collides with atoms on the surface of the food.

It is just on the outside inch or so of a food item where the heat is generated; the centre warms up only through the slower process of conduction of heat, after it has been generated. This means that the outside can become very hot while the inside is still cold. If the food is, for example, processed meat and the microwave is used for primary cooking rather than reheating, dangers can result. By their very nature of cooking, microwaves produce rapid and uneven heating, unless the item is small. The fundamental rule for using microwaves is not to rely on them for primary cooking of large pieces of meat.

Some suggested uses of microwave ovens

smallish portions of green vegetables, carrots, onions
smallish portions of fish
reheating plated meals (within 2 hours) for latecomers
thawing frozen bread
making soups, sauces and drinks from powdered
  ingredients

Doubtful uses of microwaves

cooking jacket potatoes (I simply do not like them)
pastry items
pizzas
sausages, bacon
eggs – yes, I know you can make scrambled egg this way,
  but is there any point?
large joints of meat
any food where thorough cooking is needed for
  safety

Chemical safety of microwave ovens

There have been continuing worries over the safety of microwave ovens. Not just the question as to whether contaminating bacteria are killed, but the fear for long-term human safety. Three pieces of evidence raise these questions.

Work in the USA has shown that men whose jobs involve them with electromagnetic radiation have an abnormally high incidence of cancer of the male breast. Electromagnetic radiation is the name given to invisible rays associated with high voltage power lines, telephone communications and microwave ovens. The fear over microwaves escaping from the cabinet has prompted manufacturers to ensure that the doors fit well and that any microwaves hitting the door and sides, floor or ceiling of the oven are bounced back inside, so that all the waves end up in the food.

It is not known exactly how microwaves produce heat, but this is best viewed as a by-product of the interaction between certain chemicals and the microwaves. The collision results in the energy of the wave being dissipated and the formation of some type of chemical change in the food. It is interesting that treating foods with gamma rays also produces chemical changes. Gamma rays are also a type of electromagnetic radiation and some heat, too, is generated after the food has been so treated.

It is because the identity of these chemicals produced by microwaves is largely unknown that there is concern. Microwaves have a particular tendency to react with salt and monosodium glutamate. This explains why microwaves penetrate so badly into convenience meals. But what is not known is the details of the chemicals produced after this interaction.

Work in Vienna has shown that the chemical structure of some amino acids can be altered by microwaves. Amino acids are the building blocks of proteins and enzymes, and what appears to happen is that 'mirror image' chemicals can be produced.

The point is this: microwave ovens have not been adequately researched. If we are worried about the waves escaping and reacting directly with people, should we not also be concerned about the effect of eating new chemical substances?

If there is a risk with microwave ovens, any ill effects are likely to be in the long term, and will depend on the extent of their use; they would be expected to be most hazardous for children and pregnant women.

There is another side to this debate. That is, there is no evidence of actual harm, that microwave ovens have been used in many countries for many years, and that no-one until recently has raised safety worries. I am sure the spokespersons for the UK and USA governments will say they are quite safe. They may be right: let's hope so. But this is yet another example of problems arising because new food technology has not been adequately researched.

Combined microwave ovens

People have become aware that the texture, colour and general feel of some microwaved food, although produced rapidly, is unsatisfactory. Hence the arrival of the combined oven, with a rapid microwave unit and another element, typically giving off radiant heat. More complicated, more to go wrong, more confusion?

There are some people who believe that because microwaves do not often produce colour change on the surface of food, they heat from the inside out. This is definitely not true, and on no account should you use the combination oven believing that the microwave will do the inside, and the radiant element the outside.

But does not the availability of these dual models suggest just how unsatisfactory microwave ovens are for certain types of food?

Modern technology

As an example of the problems of new technology, this recently happened to me.

I knew something was wrong, when nearing home next door's burglar alarm was screeching. Then, on going into my kitchen, the digital clock on the rarely used microwave was flashing idiotically. So was the clock on the electric cooker timer. Yes, you've probably guessed: there had been a power cut, and some malfunction had provoked the neighbour's burglar alarm into action when the power was restored.

Then the cooker alarm bell went off. There were five different control buttons, each identified only by an incomprehensible sign. I tried all possible combinations to reset the timer and hence get the oven to work at all. Always in vain; the piercing jangling always returned. So the only course of action was to switch off the power supply to the cooker. But where was it? There was no switch on the cooker or on the wall. In desperation, I switched off at the mains. At last the cooker switch was located in the corner of a cupboard hidden by saucepans. Next, to try and find the instruction book to reset the clock. After an hour I gave up, and switched the cooker off once more.

Then I wondered: why did the cooker have a clock, alarm and timer, etc? Gas cookers usually don't. Surely the idea of putting a meal in the oven all day at room temperature in preparation for a brief warming up before arrival home can be manifestly dangerous?

Is this not an example of modern technology running completely out of control? Presumably the more controls there are, the better the sales talk and the greater the value of goods you think you are purchasing. OK, the expert electronic boffins can cope with all the gadgets and they may not need instruction books. But many people, certainly including me, cannot.

And surely the more controls there are, the more there are to break down? Have you ever tried getting a cooker clock repaired? You may be forced into buying a whole new cooker. Cookers used to be a life-long investment. Now they seem yet another ephemeral consumable, with new improved models becoming available every year or so.

(It is not just kitchen equipment. Take the car radio. Many are too complicated. How many people struggle with the controls while driving, in a desperate bid to get the programme they want after someone else – no names mentioned – has reset it? Indeed, how many accidents are caused by this?)

Tips on kitchen hygiene

First, there is you

Handwashing at all times is the single most important action to take. Scrub under those finger nails with a brush that is then allowed to dry. The most important times to wash your hands are after using the toilet, after handling any raw red meat, poultry or fish, after disposing of remnants of food into waste bins and after cleaning surfaces or sinks. Ideally finger nails should not be long.

Keep a towel just for hands and change it daily and, if you can use just one sink or basin for handwashing.

Of less importance is controlling your hair, covering up spots, and so on.

But if you suffer from diarrhoea, it really is important either to delegate cooking or food preparation to someone else, or, if you can't, to wash and scrub your hands thoroughly after going to the toilet.

Second, there is the kitchen

In the kitchen, the key action is the total separation of raw food and cooked food all the time. Already we have seen that in the refrigerator these should be separated, as they should on work surfaces and different utensils, boards used for either the cooked or the raw category of food. If you have not the facilities, manipulate the cooked food first if you can or, failing that, really clean all contact surfaces and cutlery between the raw and cooked.

Like the hand towel, tea towels and dishcloths should be changed daily and kept as dry as possible.

Pet food and dishes should be kept distinctly apart from human food. There is a small risk that diseases can spread to us from the excreta of pets, so after handling pets, their food or their trays, wash your hands vigorously.

Ensure that remains of food on slicers, graters and processing equipment are physically removed after their use. Repair any crevices, such as around sinks where food remains might lurk and bacteria multiply.

Take precautions against access by our insect friends – flies, wasps, beetles and so on.

Do not let children, pets or insects have access to waste (trash) bins.

Cover any food left for longer than a few minutes, so that the risk of flies transferring unspeakable material onto it is minimised.

If spills of packeted dry food occur, remove them because they are bound to attract pests and vermin.

Finally, look after the general structure of the kitchen with regard to repairs. Happy cooking!

Antiseptics, disinfectants – what a waste!

People are wasting millions of pounds and dollars on antiseptics and disinfectants. For example, the use of creams containing antiseptic agents for applying to superficial cuts, bites and grazes of the skin are of little value, and could be harmful. For these lacerations, a thorough wash with a jet of cold water should be used to dislodge any dirt. If the skin is badly soiled, a wash with soap should suffice. Then the area should be dried thoroughly with a clean towel or cloth and covered with a loosely applied protective plaster.

However, it is the massive waste of disinfectants in the home that particularly needs addressing. Adverts stating that a bleach product 'kills bacteria dead for longer' are ridiculous. Perhaps the greatest waste is in the kitchen, on floors and work surfaces. The way to get rid of kitchen dirt is to use a detergent and hot water, with scrubbing if necessary and thorough drying. If food debris is in crevices there is absolutely no point in anointing them with disinfectant because it probably won't penetrate and the dirt could well inactivate the chemical anyway.

In particular, there is no need to wash your hands in antiseptics. At best it will waste money; at worse it will cause allergic skin reactions. For handwashing, ideally, use bar soap and a scrubbing brush to clean any dirt from under the nails or from any crevices on the hands. Afterwards, the hands should be dried on a clean towel. It is important to keep bar soap dry when not in use because if it is constantly wet, it could enable bacteria to grow on it. It may be worth having two sets of soap and holders that ensure that the underside of the soap dries out after use, and then use the soaps in rotation.

It follows from this that liquid soap dispensers, however appealing their perfume, are not recommended as after some weeks in place, you could be washing your hands in a fluid teeming with invisible bacteria.

The key to cleanliness is the physical removal of dirt and thorough drying.

The traditional view of 'airing' materials outside does indeed have a sound scientific base. The reason why we do not live in a cloud of atmospheric bacteria is that they succumb rapidly outside because of drying and damage by ultraviolet light that can even pass through clouds.

Perhaps we can identify an occasional use for simple bleach – mainly as a stain remover in toilets or in the kitchen. Otherwise it is a matter of thorough washing, scrubbing and drying.

PART V

The ideal diet

15
The ideal diet

We must first appreciate that most of us are fortunate to be able to argue about what we should be eating. Many peoples in the world are grateful for any food at all. In the developed world we have overcome many diseases that would otherwise reduce our life expectation. In general our sanitation is good. Young boys can expect *on average* to live to be threescore years and ten, and girls a few years longer.

Not everyone, however, achieves this average of what we call the natural human lifespan, and many of the residuum of diseases that have not been controlled are related to what we eat. It is emphatically not the purpose of this book to advocate a highly clinical diet for everyone. Rather, it is hoped that we can combine enjoyment of our food with excellent nourishment and safety, and with minimal damage to the environment. Ten apples, six slices of wholemeal bread, and two sardines a day may well be an ideal diet in theory, but are not practical for many people, some of whom would rather starve than be confined to this regime! There is still an active and continuing debate about the ideal diet. Not all the experts agree on everything. Some must be wrong on some of the issues! I will try and give both sides of the argument, but you, the reader, are the final judge.

Let us look initially at certain groups of people for whom diet is all important. Sometimes this importance is based on firm medical evidence, but at other times there is a large psychological factor (this is very real to the individual and if a person firmly believes that a food upsets him or her, then it is reasonable to avoid it).

The first category, which probably accounts for less than 2%

of all people, is where the sufferer has one of several medical conditions requiring a special diet. These include coeliac disease, which presents in childhood with gastrointestinal symptoms that result from intolerance to the wheat protein, gluten. A lifelong gluten-free diet is the solution. Phenylketonuria is another inherited disease, in which the amino acid phenylalanine, one of the components of proteins, has to be kept to a minimum. People who are intolerant to any synthetic chemical, such as tartrazine, should not eat it, although to be confident of avoiding this can be difficult.

It is notable that some of the very restrictive diets for treating certain medical conditions have been relaxed in recent years. Diabetics can now eat more freely than they used to: what matters is the relationship between the make-up of the diet and their drug treatment. Patients with stomach ulcers used to be given dreadful concoctions of rice pudding and disintegrating boiled fish. It is now thought that stomach ulcers are due, certainly in part, to a nasty little germ called *Helicobacter pylori*, which is treated with antibiotics, not rice pudding!

For most other people, the differences between eating the right and the wrong diet can change life expectancy by 1 or 2 or perhaps even up to 10 years. One of the problems is to be able to predict whether the diet of any one person is going to make a difference to his health. The reason for this is that there is a variable and unpredictable capacity of some people to 'handle' and eliminate food components that might be damaging in other people. Because these bodily functions are the result of the quota of genes inherited at birth, an important clue is the state of health and longevity of one's parents and grandparents. If only we could choose our parents!

Calories

Given that most people eat a varied diet, the greatest single problem is eating the correct amount of food – usually not eating or drinking too much! The amount we eat is based largely on habit, but it is quite clear that overweight children of overweight parents result from all the family eating too much.

People vary in their ability to use food; some of us are more efficient at absorbing it, some are better at burning off excess calories through the activity of 'brown fat'. This means that for any group of people performing similar physical exercise, the amount of food each person needs does vary. It is certainly possible that an overweight person might have been eating the same diet as a person of ideal weight!

The message really is very simple: being overweight means that the amount of food that has been eaten is more than the person needs. True, a number of rare glandular disorders can cause obesity. Myxoedema causes all your body functions to slow down and can make you swell up. The symptoms of this disease are weight gain, swollen legs, husky voice, dry skin, lethargy, depression and loss of hair. But even here, the weight gain is due to eating more food than is actually needed, in addition to some fluid retention. On first starting the oral contraceptive pill, women can also gain weight, partly from fluid retention, and probably also from increased appetite.

Slimming diets

I have a theory that the reason why most slimming methods don't work is that they involve eating. Let me explain.

We are accustomed to believe that if doing a little of something produces a small benefit, then doing a lot of it will cause even more. So if we work 2 hours for a wage of £5 or 10 dollars we would expect to receive £50 or 100 dollars for 20 hours.

By thinking that we can lose weight by eating, then does not our subconscious tell us that if eating one low-calorie biscuit is a good thing, then eating 10 must be ever so much better. It is true that the labels on slimming foods tell us that they should be eaten as part of a generally controlled diet, but we take notice of labels only when we want to.

The other problem of trying to slim by eating is that we are constantly thinking about what next to eat, and what not to eat. Our preoccupation with guzzling is therefore actually reinforced by trying to slim and this must explain why some people actually gain weight when on a diet. It is almost as if the slimming food is eaten in addition to the usual diet.

Slimming diets are a multi-million pound industry that helps only a minority of people. Certainly some groups, such as 'Weight Watchers' in the UK, provide valuable education and support, but may I suggest an alternative approach for those people who have been overeating for some time, and not endowed with adequate brown fat?

This is to change our lifestyle so that we stop thinking of eating, except for a main meal once a day, and for which we put a great deal of energy into its perfection. We should avoid the effortless convenience meals, even though the number of calories is stated on the packet.

An active hobby or voluntary work could be a useful diversion. To me, exercise is a very boring way of removing excess weight, and has the added disadvantage that it can induce extreme hunger, such as after swimming. I am not promising miracles by suggesting you forget about eating during most of the day, but it might work and you'll save enough money to be able to afford the highest quality food!

But the commercial exploitation of slimming is seen at its most extreme in the USA. The waste of money and resources is incredible. You do not eat to lose weight. You stop eating. What about all the magazines on slimming? Overweight Americans used to be jolly. Now they feel guilty!

But statistics have shown that despite obesity increasing in both the UK and the USA, premature heart disease has fallen in the USA over the last two decades. This is good news, and I attribute this to education and the change in the basic diet, not to the slimming industry!

Fibre

What is fibre? Most people see fibre as the stringy hairs on the outside of coconuts! Not a bad starting point. It is certainly of vegetable origin, tends to be stringy or fibrous in texture, feels dry and is often brown-coloured. But in food it may not be obvious, and tends to consist of the tiny strands of insoluble carbohydrate that the body does not digest at all easily, although some of the bacteria in the colon can break parts of it up into small pieces.

Fibre is not really a nutrient, but more of a natural laxative. After eating fibre, it increases the volume of the gut contents, which reflexly causes the gut wall to contract and propel the contents along. The result is that the weight of faeces from someone on a high-fibre diet goes up, and the interval between eating food and the discharge of the undigested remains – known as the transit time – goes down.

A short transit time (as also in airports) is thought to be good for you, and the explanation for this is that by reducing the time of contact between certain chemicals in the intestine and the lining of the gut, and therefore the amount of irritation concerned, the risk of bowel cancer is reduced. Whilst the property that makes fibre what it is – its resistance to digestion by our intestinal enzymes – is general, there are many different types of fibre. Some are made of long chains of sugars, such as cellulose. Then there is hemicellulose, pectin and gum. Lignin, as the name suggests is from trees and not made of sugars. The description of these chemicals as chains is very appropriate because each big molecule is made up of a large number of small units joined to each other.

Fibres have been in vogue for nearly 25 years, largely as a result of the work of the pioneer Denis Burkitt in Africa. Below is a list of the diseases thought to be due to a shortage of fibre, and so prevented or ameliorated by lots of fibre.

coronary heart disease	constipation
diabetes mellitus	irritable bowel disease
obesity	colonic diverticulosis
gallstones	colonic polyps and cancer
hiatus hernia	appendicitis
varicose veins	haemorrhoids (piles)
large bowel disorders	

What is certain about fibre is that because it is in effect a laxative, it can reduce constipation and in countries where fibre is a major component of the diet, the incidence of bowel cancer tends to be lower than in people who eat less of it. Diverticulitis is another bowel disease associated with constipation, and seemingly helped by high-fibre diets. The symptoms of diverticulitis and cancer of the bowel tend to be similar: bouts of colicky pain, bloody diarrhoea or constipation, and weight loss.

With the other listed diseases, the evidence of benefit from fibre is much less certain: experiments are almost impossible to perform for practical or ethical reasons. And if people do begin to eat a diet high in fibre they will have changed their food in several other respects, notably from meat to vegetables and bread. Suppose this change is associated with reduced risk of heart disease, can we give fibre the credit? No, unfortunately, not necessarily. The benefit could be due to not eating so much saturated fat or other substances in the meat, or due to beneficial effects of other substances in the high-fibre diet, such as Vitamin E. The conclusion seems to be that fibre, in at least moderate amounts, can do little harm and it almost certainly is beneficial in reducing bowel disease, and it might even have protective effects against other diseases.

What foods are high in fibre? Essentially, fibre is old-fashioned roughage and much of the fibre we eat comes from the outside layers of cereal crops, fruit and vegetables. Many processed cereals are fortified by fibre extracted during the milling of wheat to produce white flour. Believing that a moderate intake of fibre is desirable, it is apparent that eating white bread and high-fibre breakfast cereals will have a similar effect as wholemeal bread and unfortified cereals!

Sugar

Sugar has been one of the most contentious issues for a generation, and ignorance over whether we are eating too much of it, or possibly not getting enough, has polarised most observers into extreme views. The argument over sugar seems less important than other issues, and of course if we eat too much sugar, we will take in too many calories and if we don't burn it off, it will be converted to fat and stored. This also applies to proteins and polyunsaturated fats. The only aspect of sugar (sucrose) that makes it different from these is that we do not need to eat any of it at all. But the sugar pushers say it is a natural food found in fruits, vegetables and flowers. So why are so many people claiming that we are eating too much?

There are some concerns over what eating too much sugar can do to you. These are summarised as follows:

1. Sugar is usually used as an artificial sweetener and so has to be extracted from either sugar beet or sugar cane. A few foods do contain high amounts of natural sugars, such as sucrose in honey, and some lower concentrations of other sugars, exemplified by lactose in milk. But generally, sugar is a food additive and it is argued that it should have been tested for safety as such, and this has not been done.

2. Sugar in the pure refined state is very easily absorbed after eating, and is transported in the blood as glucose. The sudden rise in the amount of glucose in the blood soon after eating in turn produces an outpouring of the hormone insulin from the pancreas, which is needed to shift the glucose to the inside of the cells in the body, where it is used to produce energy or to make other chemicals, including fat. You can try the following experiment.

Drink 4 oz. raw sugar dissolved in suitably flavoured water and on another occasion eat the equivalent carbohydrate as starches

– say 1 lb. boiled potatoes. The first difference you will notice is that the potatoes will make you feel full for longer. You might well experience hunger around 2 hours after the sugar. This is because the body has slightly overdone the amount of insulin it has released and the blood glucose has dropped to such a low level that it induces a feeling of hunger.

In contrast, the starch in the potatoes will slowly be converted to glucose, the concentration of which will rise in the blood steadily and produce a less dramatic but more sustained insulin response. There should be little tendency for the 'rebound' drop of the blood glucose.

This means that refined sugars are less satisfying than starch, and increase the appetite. Is this not the reason that you feel hungry soon after eating Chinese food, because of the large amount of sugar in the sweet and sour dishes?

Ultimately the pancreas can become exhausted of insulin after years of sugar stimulation, with the end result of diabetes. This association has not been definitely proven as cause and effect.

3. The sugar may be dissolved in the mouth's saliva and provide the basic nutrient for the bacteria causing tooth decay. This results in acid produced from the sugar as a result of the enzymes, in particular by one bacterium called *Streptococcus mutans*. The counter-argument is that starches could also be used by these bacteria to initiate or aggravate tooth decay. The main reason for improvement in the condition of our teeth in recent years has been due to the use of fluoride, rather than a drop in the consumption of sugar.

Let us return to the view that states that sugar is natural and it is also part of every cell of the body. Yes, agreed, it is indeed essential for life. This is only important up to a point, because our body cells can actually make glucose from many other substances, and surely it is not 'natural' to eat a can of baked beans or chocolate cookie sweetened with cane sugar? The word 'natural' can mean almost anything the advertising people want it to mean. The all important questions are how we eat sugar, and the amount we consume.

Sweets were first developed as an occasional luxury and, like so many things in life, luxuries soon become necessities and we must now be eating much more refined or extracted sugar than is good for us. But let me make it clear; if you dip a spoon into a jar of sugar and eat the sugar there is absolutely no proof that these grains of sugar are themselves toxic to you! In conclusion, there seems to be nothing inherently dangerous in eating small to moderate amounts of sugar, but we do not actually need to eat refined sugar at all. Maybe one reason why so many experts don't like sugar these days, is that they don't like the extent of food processing? They may well have a point.

Cholesterol

There is confusion over cholesterol. The following attempts to explain what goes on in the body, but the whole story is, unfortunately, more complicated.

Much of the pioneering work was done in the USA – the series of studies known as the Framingham project – where the association between diet, cholesterol, blood pressure and life expectancy was studied over many years. There is absolutely no doubt that the higher your blood cholesterol, the greater the risk of heart attacks. This *association* between a raised cholesterol in the blood and a risk of getting a heart attack *is* established. Indeed, a high blood cholesterol is the single most important predictor of a heart attack. The issues in question are not to challenge this first-rate research, but to ask how and why this occurs and, in particular, to ask: does the cholesterol itself cause the heart attack, and does lowering your cholesterol in middle age do any good? Cholesterol is a large chemical made up mainly of carbon and hydrogen; it does not dissolve in water and can produce crystals with sharp edges capable of 'damaging' blood vessels such as those in the heart.

The trouble with cholesterol is that we do need it, both to make structures in all our body cells and to make hormones. Most of our cholesterol is made in the liver, and carried round the body in the blood to the cells where it is used. When in the blood, the cholesterol is enclosed in proteins that make most of

Cholesterol crystals

it dissolve. Any unwanted cholesterol in the blood is returned to the liver, and it is usually carried out of the body dissolved in the bile entering our intestines.

If healthy people eat food rich in cholesterol, all that usually happens is that their own production of it is reduced so the amount in the blood does not increase, at least not in the short term. A few years ago we fed 'volunteers' food rich in cholesterol, that is four hard-boiled egg yolks, and found that over the next day there was no change in the amount of cholesterol in their blood. Unfortunately, a small number of people are unable to carry cholesterol normally in the blood, and if these people eat food high in cholesterol, it may accumulate with damage to blood vessels. Blood tests can be done to see if you have this problem; alternatively, if your parents and grandparents lived to a ripe old age you are unlikely to be in the so-called risk category.

For reasons not fully understood, the fats in meat or butter (there is little polyunsaturated fat in these) tend to increase blood cholesterol in many of us. So whilst both butter and meats are not that high in cholesterol itself, it is reasonable dietary advice to eat a considerable portion of fat that is polyunsaturated. Such desirable foods include margarines and oils made with mostly vegetable oils, and fish and vegetables as a whole. Polyunsaturated fats, in addition to lowering blood cholesterol, are also needed for other body functions and are described as essential. But it must be appreciated that the hydrogenated oils used to stiffen up the oils in margarine have a potentially undesirable effect by raising blood cholesterol (see Chapter 7).

This does not mean that dairy products should be avoided altogether. Most of us can eat diets containing both saturated and polyunsaturated fats, and probably more important than the exact balance of these foods is the total amount of it. So excessive total fat intake may be as harmful as smoking as far as cholesterol and heart disease are concerned. Dietary fibre and moderate amounts of alcohol seem beneficial. The alcohol may act simply by helping to dissolve cholesterol in the blood.

But there is another side to the cholesterol story. A study has been conducted in Finland, where 1200 middle-aged men with fairly self-indulgent lifestyles were studied for 15 years. One half of these continued as before. The other half were told to take more exercise, eat more fish and margarine, not to smoke or drink to any extent and have regular medical checks for cholesterol.

The results? Oh dear, the health freaks had more heart attacks than the controls!

But before we all give up trying to eat healthily, let's look again at this study. The first serious concern is its design. Too many different factors were compared at the same time, so we do not know which one was responsible for the bad results. Was it giving up the beneficial effect of alcohol? Was it the stress, associated with keeping to the regime with all the harrassing medical check-ups and enforced exercise?

We can't tell, but the other problem with this study is that it was performed too late. We do know that heart attacks are caused

essentially by two events. The first is the roughening, followed
by furring up, of arteries in the heart that begins in childhood.
The second is the formation of clots of blood in the damaged
arteries: this occurs in middle and old age. So the study in Finland
would relate to the second part of the disease process – the
tendency for blood to clot – rather than the roughening and
cholesterol deposition that starts in young people of both sexes.

So our basic dietary advice remains, certainly for children and
youngish adults. Vegetables and their oils, fish, rice and
wholemeal bread are positively good for our hearts. Small–
moderate amounts of cheese, butter and meat and eggs are not
harmful to most people and genuinely free-range products must
be better than intensively reared. Excessive calorie intake,
whether from sugar, beer, flour, chocolate or crisps, is not good
for us.

The question of margarine is still not resolved, and it is note-
worthy that the people in Finland on margarine diets fared worse
than those eating butter. The problem with margarine is that
whilst its basic ingredient is often a desirable vegetable oil rich
in polyunsaturates such as sunflower, in order to make it stiff it
has to be hydrogenated. The resultant chemicals are distinctly
unnatural and can cause the amount of cholesterol in the blood
to rise. Perhaps we should cook with, say, sunflower oil and eat
butter in moderation?

So what advice should be given to those people who are
middle-aged with 'unhealthy' lifestyles? May I suggest, first, be
sensible with the speed of any changes you make, and do not
worry too much – it's largely too late. That smoking cigarettes at
all is bad for you is not in question. There is no need to participate
in the national American neurosis, cholesterol-phobia. The main
reason to reduce your eating of eggs is to reduce calories and
fat, rather than cholesterol.

Or you can look at it another way. If your fate is sealed by
your diet before you are aged 40, you might like to believe that
you can indulge yourself from that age. Incidentally, the evidence
that exercise prolongs life is disputed, although it can make the
heart beat more efficiently. Cynics argue that exercise only makes
you fit to do even more exercise! The word 'fit' is a verb, not a

noun. You cannot be just 'fit', you can only be fit *for* something!

Who should have their cholesterol measured?

Because of the real uncertainty over whether lowering a so-called raised cholesterol (the experts cannot agree on what level is too high) does improve health, it must be only in the people with severe cholesterol disease types that measuring cholesterol is worthwhile. In particular, if your four grandparents lived to be over 70, or they died from causes other than heart attacks or strokes around this age, then the chances of you having a cholesterol problem are small indeed. There is a huge medical and pharmaceutical industry at present exploiting people's fears over cholesterol, for their profit. In general, eat a correct diet and ignore your cholesterol unless you do have a family history of cholesterol illness or your doctor advises otherwise. Even if you do lower your cholesterol, there is real doubt that this will do any good except in instances where the cholesterol is severely raised. Anyway, it may be too late! So eat the right food and stop worrying.

The conclusion, as in most matters of health, is that all of us should aim at stopping our blood cholesterol from rising in the first place, rather than rely on drugs or drastic action to lower it. In the long term, some of these drugs might have some very nasty side effects. Will it not be a tragedy to take a drug that turns out to be dangerous for a health condition that was not helped by it? There is a distinct possibility that some drugs that are now used to lower the blood cholesterol might not improve the quality or length of life.

Fats

We are now in a position to make some comments on fats. Most of our diets contain 30–40% fat, much of it saturated. Fat in food has three functions. One is to provide energy and there is really little difference between animal fat and vegetable fat for this purpose. However, in the process of being distributed around the body, animal fat may tend to raise the cholesterol in the blood. The second function of fat is as a store for energy, were we to face food shortages – quite unthinkable for most of us. The third function is as an essential nutrient: we do need some polyunsaturated fats. The mono-unsaturated fats, such as olive oil, seem to be associated with lower blood cholesterol levels than saturated animal fat, and they are excellent sources of energy, and are not associated with adverse properties.

I do believe that dieticians are spot on when suggesting that the following changes in our diet are rational.

1   Our total fat intake as a percentage of calories from food should be not more than 25%.
2   This should include mono-unsaturated and polyunsaturated but some saturates cannot be harmful. The question of the safety of hydrogenated oils is very real.

Vitamins

What are vitamins? These are types of chemicals the body needs to function properly, but it cannot make them from other substances. For example, the cells in the body can make sugar (glucose) from protein, but they cannot make ascorbic acid (Vitamin C) from anything. So Vitamin C is a vitamin. Sugar is not.

Why can't the body make vitamins? This is best understood by appreciating that as organisms evolved they became better at doing some things, but at a price. Their cells, like the UK, have lost some of their manufacturing ability! This means that most vitamins are made by primitive organisms, and overall the more primitive they are, the better their ability to make vitamins. Plant food is, therefore, a superior source of some vitamins, such as folic acid, than meat. However, vitamins can be present in animals as a result of their eating plant food.

Bacteria and yeasts are even more capable. Some bacteria are able to live entirely off water, dissolved mineral salts, and gases from the atmosphere, and can convert gases in the air, such as carbon dioxide and nitrogen, to substances as complicated as DNA.

The traditional exhortation to growing kids, 'Eat up your greens', does have a sound scientific basis for the provision of vitamins. We can see, once again, the significance of man's vegetarian ancestors. But, nothing is ever quite as simple as it appears at first.

There is one vitamin which may be deficient in a strict vegan's diet and that is Vitamin B_{12}. This is made by bacteria (including the bacteria in our own intestines) rather than animals or plants. So what is the problem? Unfortunately, the B_{12} that the bacteria in our intestines make is not absorbed adequately from the lower bowel. We need a substance in our stomach (called intrinsic factor) to absorb B_{12} from food. B_{12} is found in milk, cheese and eggs, so the conventional vegetarian should not run into difficulties, although vegans might need occasional supplements.

Incidentally, I do not accept the argument that bacterial contamination of food is generally desirable because of the formation of B_{12}!

Vitamins E and C – the good guys

You've probably noticed that the labels of much of our processed food mention antioxidants, including ascorbic acid (this is the same as Vitamin C), and chemicals with 'E numbers', for example E300–E312. Antioxidants are added to stop food decomposing as a result of oxygen in the atmosphere reacting chemically with some of the constituents. It is important to stop this for two reasons: the new chemicals produced might not taste at all pleasant, such as in rancid butter, or some of the products might actually be dangerous. So the less often you shop, the longer processed food has to last, and the more antioxidants are needed (see Chapter 7).

The problem of oxidation, as the chemical process is called, occurs not just in food, but also in our bodies, and one long-term damaging effect of this is thought to be cancer. About three quarters of all cancers are due to environmental factors acting on individuals who are vulnerable to the disease. (For example, one in nine of heavy smokers develops lung cancer, but not everyone.) Of these cancers about half are thought to be due to eating the 'wrong' food.

There is some research that suggests that if we do not eat enough natural antioxidants, then we may be more vulnerable to cancer, years later. There have also been suggestions that the benefits of natural antioxidants extend to preventing heart attacks.

Two well known vitamins, E and C, are natural antioxidants, and the research tends to show that in people with various types of cancer, the amount of these vitamins is less than in people who do not suffer from cancer. The theory would suggest that chemicals in the body produced by oxidation irritate and damage cells with the ultimate production of cell overgrowth, or cancer – in the same way that smoking causes lung cancer. The natural antioxidants prevent this process, and of course with a tendency to favour stale processed food, these antioxidants tend to get used up in chemical reactions in the food before we eat it.

So what does this theory, (and as yet there is no definite proof), mean for what we should eat? The first suggestion is that fresh food is generally preferable to stored food. There are some herbs with potent antioxidant effects. These include cloves, mace and oregano, but you can't live on these alone!

It is foods high in vitamins C and E that provide most effects in practice. Vitamin C is found in fresh fruits, particularly oranges and lemons, fresh vegetables and potatoes. Cooking food to high temperatures and storing food for too long results in loss of this vitamin. So, sorry to have to say this, the pork pies and mushy peas so popular in parts of the UK would be just about the worst meal for Vitamin C content.

Vitamin E (also called tocopherol) is found in many foods, including wholemeal bread and vegetable oils. There are unquestionable dangers from eating too much of some vitamins, such as A and D, but not apparently with vitamins C and E.

B Vitamins

B vitamins are the biggest group of these essential trace chemicals, without which we are liable to decline, develop illnesses and even depart from the world. But there are many misconceptions. Some so-called B vitamins are not really needed at all and therefore are not actually vitamins. There is, for example, no proven medical benefit from eating the substance choline, despite numerous claims by the distributors of this chemical.

Other B vitamins are definitely essential and it is possible to calculate a daily requirement for these. But this does not mean that we actually need to eat the exact dose of these vitamins – e.g. thiamin (B_1), riboflavin (B_2), pyridoxine (B_6) – every day. The body can store them for quite long periods. For example, B_{12}, which is called cyanocobalamine, or just cobalamine, needed for prevention of anaemia and nervous malfunction, can be stored for months or years in the human liver. People who are unable to absorb this vitamin from the diet, and so suffer from pernicious anaemia, can be treated by an injection of B_{12} as infrequently as once a year.

It might be better, therefore, to look at the intake of B vitamins (and others, for that matter) on a weekly or a monthly basis, rather than daily. The great majority of diets will then be found to contain adequate B vitamins, when reviewed over the longer term. It is true that extra supplementation of B vitamins is unlikely to do any harm (or perhaps we should be a little more cautious and say that little research has been performed on the effect of eating lavish amounts of B vitamins, so no side effects have been identified). But the ritual self-medication of B vitamins in addition to a normal diet must be unnecessary and costly and it might actually divert you from changing from a poor diet to a better one. Claims that supplements of Vitamin B and other vitamins enhance children's intelligence remain very controversial.

Some alcoholics do suffer from deficiency of B vitamins, not because of the alcohol itself, but the lack of B vitamins in the near exclusivity of their fluid refreshment. On admission to hospital for 'drying out', patients are usually given supplements of B vitamins. This knowledge has somehow become distorted in the general field of popular dietetics and nutrition. There is a completely erroneous belief that supplements of B vitamins can neutralise the damaging effect of alcohol excess. Sorry to say that, like anything else with the diet, the response to problems of excess is to reduce the amount of substance consumed.

Folic acid is a B vitamin that requires special mention. As the name suggests, it is found in foliage, or more specifically in leaves of plants. Folic acid is needed in the body for several functions, including making new DNA in our cells when they divide to grow or replace damaged parts.

For some years, suspicions have been expressed that deficiency in folic acid might produce serious defects in the developing body, particularly during the first weeks of pregnancy. In 1980 it was shown in Leeds, UK, that women who voluntarily took vitamin supplements (including folic acid) were less likely to give birth to babies with neural tube defects (i.e. defects of the spinal cord and the brain) than patients who did not take extra vitamins.

The beneficial role of folic acid was confirmed in July 1991. This provoked a letter from the Chief Medical Officer at the Department of Health to all UK doctors, pointing out that 'folic acid supplements, taken by women who had already experienced one or more pregnancies affected by neural tube defects, reduced the risk of recurrence by 72%.' Similar advice has been provided in the USA.

The crucial time for the administration of folic acid is right at the beginning of the pregnancy, often before the woman would be certain that she was pregnant. The implications of this for our diet are substantial, as it is reasonable to require that any woman who might be 'at risk' of becoming pregnant should receive a diet adequate in folic acid. While the incidence of neural tube defects is not high (typically about one pregnancy in 200–400) the effects can be devastating as a result of brain damage or paralysis from disease of the spine.

Traditionally, pregnant women have been encouraged to ensure adequate vitamin intake through, for example, eating liver. But the Chief Medical Officer in the UK, in his letter of 12 August 1991, has pointed out that 'earlier advice stands that women who may become pregnant should not eat liver, because it may have too high a vitamin A content.' There is good reason to believe that excess Vitamin A is associated, paradoxically, with yet further congenital defects, such as those of the face. The possibility that excessive Vitamin A can be dangerous will come as a shock to American readers, who may have succumbed to the advertisements recommending Vitamin A supplements. I am not aware of data on American liver and its Vitamin A content.

The main natural sources of folic acid are vegetables and fruit, and while frozen peas still retain a substantial amount of folic acid, vegetables subjected to chilled storage may not. Thus the advice to women possibly anticipating pregnancy must be to eat adequate fresh vegetables and fruit. This is not, of course, a decisively novel approach, but recognises that many of our new types of food processing will result in food depleted of folic acid, unless specific supplements are incorporated.

The typical contents of folic acid in a variety of foods are shown below.

Food	Folic acid content (µg)
1 helping of boiled potatoes (180 g)	50
1 helping of boiled spinach (90 g)	80
1 helping of boiled green beans (90 g)	30
1 helping of boiled brussels sprouts (90 g)	100
1 helping of boiled cauliflower (90 g)	50
1 helping of boiled frozen peas (100 g)	45
1 tomato (85 g)	15
1 orange (160 g)	60
½ avocado pear (75 g)	50
1 banana (100 g)	20
½ grapefruit (80 g)	10
1 slice melon (180 g)	50

2 tablespoons bran (14 g)   40
1 helping fortified cornflakes (40 g)   100
1 helping fortified branflakes (40 g)   100
1 helping spaghetti (230 g)   10
3 slices white bread (90 g)   30
3 slices wholemeal bread (105 g)   40

Reference Nutrient Intake (RNI) for adult women – 200 µg/day

In general, the B vitamins, as a group, are found in wholemeal cereals, peas, beans, fruit, fish, yeast products and eggs.

Vitamin K

This vitamin has received some unjustified criticism recently because of the need to give supplements to babies to prevent them bleeding. The claims that injections of Vitamin K are associated with cancer later in life were based on preliminary surveys which have not been confirmed. Even if there was an association between the giving of this vitamin and a subsequent disease, the condition of prematurity (hence the need for injections) itself could be the important factor, not the Vitamin K.

Vitamin K is found widely in many foods, and if there is a problem, it is rather one of needing to lower levels of it in patients with diseases that tend to make the blood clot too early. The commonly used drug to achieve this on medical supervision at the moment is called warfarin, also used to poison rats.

Vitamin D

Vitamin D, like A and K, dissolves in fat, and is found in fish oils (not just cod, but many deep-sea fish) and other fats. It is also synthesised in the skin as a result of exposure to sunlight. A lot is known about Vitamin D: both too much and too little of it is harmful.

The main benefit from it is that it is required for normal growth of bones and teeth, and children short of it develop bandy legs (rickets). If too much Vitamin D is eaten, too much calcium can be absorbed from food in the body and this can end up in the kidneys with disastrous effects.

A long held theory is that people who inhabit tropical countries have developed a dark skin to filter out some of the sun's rays to prevent too much Vitamin D being formed in the skin. It is interesting that the darkest skins of all people are found among the inhabitants of the dry, bright equatorial regions of East Africa.

This explains why extra supplements containing Vitamin D must be carefully controlled, and it is amazing that this vitamin is readily available as tablet form in chemists and pharmacies. The main danger of this vitamin today is that of excess, not shortage.

Herbs

There has been renewed interest in medicinal herbs. One reason for this is the increasing cost of new manufactured drugs, one with a price tag of £500,000 or a million dollars an ounce!

Do herbs actually work, and if they do, how do you know how much is needed? We can view medicinal plants in three ways. There are those that unquestionably contain substances that have dramatic effects on us. Such plants are usually described as poisonous because there is always a danger of eating too much. However, these chemicals can often be used medically after purification and the identification of the correct dose. Examples of poisonous plants are foxglove, deadly nightshade, yew berries, henbane, Lords and Ladies, and among fungi, the red-and-white fly agaric is notorious. So don't eat these!

The next type are those in which a beneficial effect has been demonstrated – and there are no worries about eating too much. Folic acid in vegetables is an essential B vitamin, and seemingly cannot be eaten in excess. The seeds of the evening primrose plant are rich in a fatty acid, called gamma-linolenic acid. The clear yellow flowers open in late summer evenings on stems about 3 feet high. This fatty acid is thought to have several medical benefits, although many claims are still unproven. There is little danger from eating too much linolenic acid, as the body will break the molecule into small pieces and use it up for energy.

The rumours that garlic kills bacteria and fungi have a sound basis. We can demonstrate this easily by squeezing a clove of garlic onto bacteria in the laboratory and showing that the bacteria are killed. As to whether it can be used medically is not clear. I suppose it could be worth trying rubbing a cut clove over an infected spot, although don't expect to be popular! After eating garlic some of it gets into perspiration, and I wonder whether the French and Italians have few skin infections because of this!

We are now moving towards the more speculative use of herbs, beautifully described by Lesley Bremness in *The Complete Book of Herbs*, published by the National Trust in the UK. I'll give you some examples, but am not promising miracles: lavender for acne, black pepper for toothache, sage for menopausal symptoms, basil for mental fatigue, lemon for colds, peppermint for flatulence, thyme for sinusitis, fennel for bruises, ginger for diarrhoea, cinnamon for insect bites, and rosemary for rheumatism. You can be assured that although these may not actually work, they will not do any harm, which cannot be said of some of the expensive and potent pharmaceutical drugs.

Wouldn't it be marvellous if the cinnamon-spiced muffins, doughnuts and buns, rightly popular in the USA, actually worked in treating some maladies!

Salt

Salt has a bad reputation. Is this justified? Yes and no is my answer. It has been very well researched by doctors studying certain patients in whom it is *known* that too much salt can be dangerous. What seems to happen is this.

In many people, the amount of salt eaten in the diet is extra for the body's needs on that occasion. We do normally eat more salt than we may need because we cannot predict how much we might be going to lose by sweating. The excess normally goes out in the urine. But in some people not quite enough salt goes through the kidneys into the urine. This has been best demonstrated in patients with heart failure, when an inappropriate reflex seems to set in and the salt is retained in the body. Salt drags with it water and both tend accumulate in the body, according to gravity. This can result in swollen ankles or legs, and is known as dropsy. This is a well known medical problem and is treated with drugs that encourage salt (and with it, water) to be lost from the body. The drugs are known as diuretics and patients with dropsy will be given only just enough salt for their needs. This is an acute medical problem and does not apply to the diet of most people.

Before we think about whether we should reduce the amount of salt eaten, it should be appreciated that too little salt in the diet accompanied by high loss through sweating or diarrhoea can be dangerous, even rarely fatal.

There are, however, some kinds of high blood pressure that result in the accumulation of too much salt in the body. For reasons not fully understood, insufficient of the superfluous salt is removed in the urine over months and years. A little extra persists in the blood and with it some additional water. There is a tendency for the volume of blood circulating around the body to increase but then another reflex comes into being: the muscles in the walls of the arteries distributing blood around the body contract and the pressure in the circulation goes up. This results in the tension in the blood rising, known as high blood pressure. Over the years, it is aggravated by high amounts of salt in the diet, and unless treated it can result in heart attacks and strokes. But as to how successful treatment really is remains unclear, as is the exact stage of the disease at different ages that merits treatment.

It is not in doubt that the problem of treating high blood pressure would not exist were the disease to be prevented in the first place. Thus in some people, too much salt in the diet is harmful.

The problem is that we cannot identify exactly which people are at risk, and two views are tenable: one is that because the great majority of people can tolerate high salt diets, then there is no need to restrict salt intake. The other is that in order to protect the unknown minority of us in early life, we should all eat less salt. The first view is that of the salt purveyors or the food industry, the second is that of doctors and nutritionists. Some experts go further than this, believing that many if not most people are eating too much salt. As with sugar, might this view be developed because of the dislike of the amount of processed food now being eaten and the fact that much of this food is supplemented with sugar and salt?

I will briefly mention another reason to reduce the amount of salt in processed food: that is, it spoils many of the natural flavours derived from natural fresh vegetable sources. Salt, like everything else we eat, is fine in moderation! I am not going to suggest a daily figure because you will not be able to calculate at all easily the amount in the diet.

Overheard in the British Ministry of Agriculture, Food and Fisheries (by repute)

The Minister, full of enthusiasm for his new job, asked 'All our experts say too many chips [that is large, fatty fries] are bad for you and eating habits are established in childhood, so I want all chips banned from schools.'

'Yes, Minister', the Civil Servant replied soothingly and under-standingly. He went on, 'Of course, banning chips is a good idea, but wouldn't people see it as a threat to their freedom of choice?'

The Minister was not to be deterred. 'If eating something is bad for you, then it should stop.'

The Civil Servant said, 'I absolutely agree in theory, like smoking, but we still condone advertising of cigarettes.'

'That is different', said the Minister, who went on, 'we have to protect the cigarette industry.'

The Civil Servant looked pained and explained condescend-ingly, 'That's the impression we have had to give, but the real reason we have to support smoking is that it reduces the numbers of pensioners, and if there were more we could not afford to pay them all.'

The Minister was beginning to get the message. 'You mean that you want people to go on eating lots of chips because they are bad for them?'

The Civil Servant seized on this statement with alacrity. 'Minister, are you saying you want this to be your policy? You will be very popular. People like chips; fatty food is easy to flavour, it will help use up some of the unwanted fat mountain, and the makers of convenience chips will be delighted.'

The Minister said 'No, we can't do that. Our advisers say too much fat is bad for you.'

The Civil Servant came to the rescue. 'Yes, Minister, we cannot promote bad food. But we can accept the findings of our advisers, issue bland statements to the public about the need for a balanced diet and then take no action, as usual.'

The Minister said, 'But don't you think we have a duty to protect the public more positively?'

'Certainly not,' said the Civil Servant, who continued, 'Suppose everyone gave up smoking, gave up chips, meat and butter and became vegetarian, they would live to be 100, before going slowly demented. Who would look after them?'

The Minister thought he had a good suggestion.

'They could always shorten their lives through drinking!'

The Civil Servant beamed ingratiatingly. 'Well done! Not many Ministers have thought of this. Why do you think that over the years we have had to raise taxes on cigarettes because of those blasted medical busybodies, but at the same time we've also reduced the real level of taxes on strong alcoholic drinks.'

Protein

How much protein do you need? Can you eat too much? Can the body produce its own protein?

The key to understanding proteins is that they contain the

element nitrogen in a form the body can use. Most of the world's nitrogen is in the atmosphere as gas and is completely useless to us. Plants, some more than others, are able to convert, or fix, this useless gas into substances needed for cell growth. Certain bacteria in the soil first convert the gas to nitrates that are then used by the plant.

In our bodies, nitrogen is found mainly in amino acids within protein, and also in nucleic acids; some of the amino acids cannot be made by us at all and these are called essential. Other amino acids can be synthesised in the body and these are termed non-essential. But to obtain these non-essential amino acids, we may to have to make them from something else. This means we do need to obtain a certain amount of non-essential amino acids from protein, even though it may not be that important exactly which.

For the typical adult, 10–20 g of protein daily, that contains the essential amino acids, is probably enough. This is the amount found in several slices of bread, an egg, a few ounces of cheese, beans, meat or fish. Young children and pregnant women need more, but not that much more.

The ideal diet in practice

Below is the World Health Organization's recommendation in 1988 of the make-up of the ideal diet for Europeans.

Ingredient	Amount (as % of energy) or weight per day
Total fat	20–30%
Saturated fat	10%
Polyunsaturated fat	About 10%
Cholesterol	Less than 250 mg (about 100 g butter)
Sugar (sucrose)	10%
Salt	5 g
Starches	45–50%
Fibre	More than 30 g

Various authorities have tried to simplify the range of foods available and categorised them as healthy or unhealthy. This is helpful, but it seems to give a false impression of the inflexibility of our digestion.

It is not disputed that vegetables, potatoes, fruit, salads, fish and wholemeal bread are good for you. So are most types of rice, nuts, olive oil, and real free range eggs in moderation.

Lean red meat, skinned poultry, hard cheese and crisps are not quite so good for you, and fatty meat, burgers, sausages, butter, soft cheese, biscuits and chocolate in excess are bad for you.

But other factors have to be considered. Some desirable foods may either not be available or be too expensive, or you simply don't like them! Do we really have to have a salad garnish three times a day, or six apples, and is white bread totally *verboten*? No, of course not, many of these ritualistic rules are over the top, and not meant to be taken too literally; they are simply aimed at making you change your diet just a little.

In addition to being the basis for health, eating does provide pleasure and is an essential part of our social fabric; and as stressed throughout this book, environmental considerations must receive more of our attention, as I am sure will the issue of animal cruelty.

So here are some suggestions, all of which could provide be a satisfactory day's eating and drinking. In general the volumes of the food are not stated because it is up to you to eat the amount that is appropriate for you. We look at this from a different angle in Chapter 17. I am intent on presenting a feel, or understanding, of the issues, rather than imposing dogmatic rights and wrongs. In any case people do vary in their needs. We cannot be titrated like a chemical reaction.

OPTION 1
   Breakfast   High-fibre cereal
                           Skimmed milk
                           Dusting of sugar
                           Fruit juice
                           Coffee

Lunch (light)	Freshly baked white French bread
	Small helping of butter and soft cheese
	Glass of wine or beer (or water!)

Supper	Fruit cocktail
	Side salad of tomatoes, lettuce, nuts and
	apples and olive oil dressing
	Steak (not from the UK)
	Jacket potatoes
	Chocolate éclair

OPTION 2

Breakfast	Bowl of sugared cereal flakes
	Full cream milk
	Coffee

Lunch (light)	Rye bread or biscuits with fish paste
	Apple and banana

Supper	Avocado pear with vinaigrette dressing
	Vegetarian egg and bean flan
	Boiled potatoes
	Gooseberry fool and a little cream
	Glass of wine

OPTION 3

Breakfast	No food
	3 cups of tea (I do not believe food at
	breakfast is necessary)

Lunch	Chicken burger, chips and salad garnish

Supper	Prawn cocktail
	Shepherds pie (minced lamb and mashed
	potato)
	Side salad

Fresh pineapple in kirsch
Coffee and petit fours

OPTION 4
Breakfast   Baked beans on wholemeal toast or
                     kippers

Mid-morning   Tuna or cheddar cheese and salad in a
                     wholemeal bread sandwich

Early afternoon   Slice of pizza

Evening   Fried egg (in sunflower oil) well cooked
                     on white toast, mashed potatoes, variety
                     of vegetables
                     Apple pie
                     Glass of beer

Babies

The pendulum at least seems to have settled down as to whether
and for how long babies should be breast-fed. It now seems as
certain as it ever will be that breast-feeding for around 6 months
is ideal.

The main advantages claimed for breast-feeding are two-
fold: one non-controversial, the other definitely so. The non-
controversial benefit of breast feeding is that the mother's milk
provides chemicals called antibodies that protect the baby from
infections, in particular gastroenteritis.

It is this benefit that has incensed many people over the practice
of producers of milk powder giving free samples to mothers just
after childbirth in the Third World. Using artificial milk soon
makes the natural flow of breast milk dry up, so on leaving hospital, the mother must continue to purchase artificial milk from the

same company, with an inevitable increased risk of gastroenteritis, and of course the cost can be prohibitive.

The more controversial claim for breast-feeding is that children who have been breast- rather than bottle-fed have a slightly higher intelligence, as judged by certain tests. It is not immediately obvious as to why this should be so; and many factors contribute to intelligence and it is possible that these, notably the genetic make-up of the parents, might be a more important factor than whether babies are breast- or bottle-fed.

There has been a survey suggesting that babies breast-fed for too long might also develop risks of certain diseases. This is why the length of time of 6 months is about right.

How many calories does a baby need? The answer is a surprisingly large amount, for two reasons. The first is that the smaller the body of an animal, the greater its heat loss, meaning a greater need for energy, and secondly, and overall most important, babies need extra nutrients for growth.

I quote the typical calorie needs from Sylvia Hull's excellent book, *Safe Cooking for a Healthy Baby*.

	Calories per day	
Age	Boys	Girls
0–3 months	545	515
4–6 months	690	645
7–9 months	825	765
10–12 months	920	865
1–3 years	1230	1165
4–6 years	1750	1545
Adult	2550	1940

Below are illustrated two of Sylvia Hull's possible healthy diets for a 1–2 year old.

Menu 1	Calories
1 pint (500 ml) milk (for whole day)	360
1 oz. (25 g) wheat flakes	100
1 boiled egg	80

Fish pie	120
2 tablespoons spinach	20
Apple brown Betty	90
1 tablespoon natural yoghurt	15
2 slim slices bread	120
¼ oz. (6 g) butter	60
1 oz. (25 g) grated Edam cheese	90
1 finger banana and date cake	80
½ satsuma	10
Apple juice, diluted	30
Total	1175

Menu 2	Calories
Quick yoghurt porridge	90
1 banana	80
Orange juice, diluted	30
Baked lentil roast	150
2 tablespoons Brussels sprouts	25
Blackcurrant yoghurt	90
1 soft wholemeat roll	140
2 teaspoons smooth peanut butter	70
1 rice and apple cake	50
1 pear, peeled, cored and quartered	40
1 pint (500 ml) milk (for whole day)	360
Total	1125

16

Hospital food – need it be so bad?

Hospital food has a bad reputation. So has school food. So has other institutional food, and all often deservedly.

There are bound to be difficulties in feeding so many people at the same time: sometimes it is quite literally 'at the same time' for, say, a hospital ward; on other occasions, there is a need to satisfy the unpredictable appetites of an unpredictable number of people passing through a canteen at unpredictable times. Very difficult, particularly if the customer requires the plated meal to be hot, but the quality unblemished by not holding it hot for long! We expect the perfect hot meal under all conditions!

One problem has arisen from the greater choice and awareness of food generally, and variety is demanded in institutions. This concerns not just ethnic food – where alternatives to, say, pork and beef have to be provided – but also to be considered are the increase of vegetarianism, and the move away from eating red meats and meat reared under intensive conditions. These needs can all be met using conventional catering methods, and if the various options available are explained to patients, they may be satisfied with two, or even one, hot meals daily.

But meals to many convalescing patients are the central event of the day, and there is no doubt that the quality is often lamentable. The explanation is glaringly obvious: institutions have not invested adequate money into provision of high quality, safe and

nutritious food. The argument often seems to go that since
patients are only briefly hospitalised, then nutrient content matters little. One catering manager has said to me that he worked
on the principle that the patient's food had no nutritional value
whatsoever! The school lunch, typified by sausages, beans and
chips (fries) has also been justified on the assumption that other
meals at home or elsewhere will make up any shortfalls in nutrients. But the school lunch is important, not just in providing
nutrients, but in establishing social patterns for the remainder of
a person's life: it is no wonder that sausages and chips (fries) are
still so popular among adults!

As mentioned elsewhere, one of our problems has been that
we have been trapped by our traditional concept that a meal
should consist of meat and two veg., hot. The Texan recovering
from his hernia operation is unlikely to be impressed by macaroni
cheese or bean curry and rice, and he will certainly want his
sausage, egg and bacon for breakfast, and steak at all other times,
however horribly they are prepared. This brings us to the point
that we will continue to receive substandard food if we are prepared to eat it.

It has always been amazing to see people travelling on planes
and trains tucking into their reheated, limp, decomposing break-
fasts. The description and picture on the menu may be appealing:
eggs, bacon, sausages, hash brown potatoes, mushrooms and tomatoes pictured as they were once.

There has been a limited change in the UK towards the preparation of these meals on the train from raw, rather than providing
reheated, precooked fare. Good news. Similarly, improvement
has occurred on some airlines in that a range of reheated items
is being replaced by cold food. But of course it is practically
impossible to produce freshly cooked food to order in the air,
particularly on short flights. In a 30 minute flight (i.e. the time
in the air), between Toronto and Detroit that I recently experienced, you should have seen the panic in trying to serve a meal
in addition to the mandatory drinks service. When the tables had
to be stored on landing, there was rubbish everywhere!

At present, the two greatest outstanding problems are school
and health care food. With both, the problems are due to lack of

investment and the preoccupation with what is most convenient for the caterer, rather than what is most appropriate for the customer.

Up to the 1960s, most hospitals in the UK and the USA produced meals through what is now called conventional or traditional methods, or as a colleague describes the system, cook–eat. This means that meals were prepared fresh, using a procedure that was essentially a scaled up version of the kitchen of a large private house or a small hotel. The two principal problems with this method in feeding, say, 500 patients, arise from the scale of the kitchen operation and the difficulty in delivering all the meals hot. Take a large vat of stew (lumps of steak cooked in a gravy and carrots and onions), for example, that may be kept warm for some hours. Suppose this is not mixed adequately and parts are kept insufficiently hot to prevent certain bacteria growing. One of these is *Clostridium perfringens* and this little troublemaker can divide every 15 minutes. This means that after an hour, one bacterium could have become 16; 2 hours, 256; 4 hours, 65,000; 6 hours, 17 million. This bacterium has produced outbreaks of diarrhoea in many hospital patients.

The difficulty in keeping food hot in transit is that it should be at sufficiently high temperature to stop bacteria growing but not too hot to spoil. A safe *minimum* holding temperature of 63 °C is now adopted fairly generally and carefully insulated trolleys are used for short journeys to keep the food above that temperature. Many of these problems with conventional cooking can be overcome with improved equipment and employing adequate staff to ensure safety. One valuable development has been automated plating devices, using a conveyor belt principle, so that plated meals are delivered hot to the ward, thus bypassing the need for nurses or other staff to plate food out from bulk containers.

During the 1960s, 1970s and early 1980s in particular, there was insufficient investment in catering in UK hospitals, and to a lesser extent in USA public hospitals. The kitchens were often old and dirty, the equipment was out-of-date, and the morale of the poorly paid staff low. Feeding patients took a very low priority, certainly when compared with the need for the latest wonder drugs. This deplorable state of hospital food – typified by luke-

warm gristle and gravy, yellow cabbage and blackened, discoloured, disintegrating potatoes – was maintained in the UK
partly by what was called 'Crown Immunity', which meant that
hospitals (and, indeed, prison kitchens) did not come under the
responsibility of local authority environmental health departments
and their health inspectors.

The outbreaks of food poisoning that did occur, not infrequently, were the responsibility of the Health Department,
which was immune from prosecution by the local department.
Few patients took it on themselves to initiate legal action against
the health authority, mainly because they knew they were dependent on them for their medical care.

This state of hospital food in the UK continued, until the
inevitable happened: a really large outbreak of food poisoning. It
occurred over a hot weekend at the end of August 1984 in Stanley
Royd Hospital, Wakefield. This hospital began its days as a Victorian lunatic asylum and the kitchen had hardly been updated
within living memory (I know, because I have been there!). In
the ensuing days, 455 patients and some staff developed food
poisoning due to *Salmonella typhimurium*. Nineteen patients died.
The media interest was intense, mainly because of the difficulty
in finding the source of the salmonella. The problem was not
helped by many of the nursing staff eating the patients' food
illegally (they should have paid for it in their own canteen), so
attempts were initially directed to finding a food common to
patients and staff.

In order to cut costs, the hospital had not employed the usual
staff responsible for tracking down infections. Outsiders were
brought in and, after weeks of argument, the likely explanation
for this outbreak was given that sliced cooked beef had been left
out overnight in the warm, to be served and eaten the next day.

The scale of this problem and the number of deaths amongst
the long-stay psychiatric patients prompted a public enquiry,
which was held in 1985. When published in early 1986, it was
damning of the catering staff, management and facilities in the
kichen. Indeed, it was critical of almost everyone except those
responsible for the level of contamination from farming practices.
In 1986, crown immunity was lost from National Health Service

hospitals, which could now be prosecuted by local authorities if they contravened the statutory hygiene regulations, or failed to implement specific action requested by Environmental Health Officers. Since 1986, the number of food poisoning incidents in hospitals has declined, but not all the consequences have been desirable.

In 1986, Wakefield Health Authority, without consultation with their catering or medical staff, endeavoured to introduce a type of cook–chill catering. The plan was as follows. A new kitchen already earmarked for the Stanley Royd site was to be converted to a cook–chill central production unit where the food was first cooked, chilled and stored. The food was then to be transported in refrigerated lorries, to several existing kitchens, where after further storage in these kitchens (after appropriate modification) the food would be reheated, transported and served, using the existing archaic procedures.

The system proposed was therefore: purchase of raw food → cook → chill → store → transport → store → reheat → transport → serve → eat (hopefully). This system was claimed to save money through using fewer catering staff for cooking, although the costs of the reheating or serving seemed not to have been considered. The plan was not as outrageous as it seems; because of the initiative of the catering arm of the Department of Health, these systems had already been introduced into many NHS hospitals. There had been no evaluation of nutrition (of course, this did not matter), nor of the safety of the system. Because of the absence of safety data, part of the proposal required that batches of food would have to be tested for bacteria to see if it was safe to eat!

Something was obviously wrong and, not surprisingly, opposition to the scheme mounted amongst staff at Wakefield; and the Yorkshire Regional Health Authority assembled a group of experts, all of whom had previously been enthusiastic for this method of food provision. As with so many failings in British food, this had been an attempt to introduce a system that had been developed in the USA for some years, but cutting financial corners, so resulting in a drop of quality and safety. The conclusion of this group was that the Wakefield proposal was perfectly acceptable as long as the delivery of the reheated food took no longer than 15 minutes! Yet in practice it would take up to 2 hours!

The NHS managers behind the scheme were soon to face humiliation, and the scheme had to be modified to involve reheating food at each ward, an extremely expensive option. This system of ward reheating was soon to be imposed on all other NHS hospitals, who were required to 'invest' a great deal of money to comply with the new requirements. The humiliation of the UK Department of Health is now complete

Apart from being hot, the quality of ward reheated cook–chill food still leaves a lot to be desired, yet the caterers have seemed determined to proceed. Perhaps the two influencing factors were that a major outbreak of salmonella food poisoning was unlikely since any defect would be likely to apply at individual ward level, and that it provided the fabric to buy in chilled food from the private sector.

This type of cook–chill food production was a development from the cook–freeze system which had been one attempt to feed institutional residents from the 1960s in the USA and the UK. In cook–freeze, the safety is rarely in doubt, because deep-freezing of stored food has a substantial margin of safety, were temperatures to rise unexpectedly. The cook–freeze system first cooked food, usually in bulk, rapidly deep-froze it, stored it frozen for up to several weeks, and then reheated and served it at ward level. Fluids – custard, gravies, sauces – would usually be prepared in local ward kitchens. The quality of reheated frozen food is average: vegetables certainly deteriorate, particularly if not eaten immediately after reheating.

The criticisms of both cook–chill and cook–freeze systems are summarised as follows:

1　The quality of the food must be expected to be inferior to that of conventional (cook–eat!) cooking. If this is not the case, the present cook–eat system needs to be improved. It may be attempting to provide food for too many patients and staff from a single kitchen. Additional kitchens may need to be established.

2   The energy used, usually electricity, must be greater for both cook–chill and cook–freeze, compared with conventional. The main advantage of cook–chill over cook–freeze is that the cost of cooling and reheating chilled food is less than frozen.

3   There has been no proper evaluation of these systems relative to conventional, as regards to all the costs, quality, safety and nutrition. There is every reason to believe that cook–chill systems will lose Vitamin C, folic acid, and polyunsaturated fatty acids to a greater degree than other systems.

4   The equipment needed for cook–chill and cook–freeze is expensive to maintain and replace.

The main stimulus for cook–chill and cook–freeze has come from the vested interest of the equipment makers, fears of a major food poisoning outbreak, and the need to put contracts out to tender. The ward reheating arrangement does make a major epidemic of food poisoning unlikely. The sad problem in the UK is that in attempting to improve institutional catering, cook–chill and cook–freeze have been considered the only way of introducing private caterers. The idea that these organisations could run cook–eat kitchens has only recently been considered.

But surely the answer to providing better hospital food is to improve conventional catering, and really evaluate the need for so many hot meals. Perhaps a traditional kitchen should be required for every 300–400 patients or staff?

The quality of food really does matter to patients, and it is patients who should be the final arbiters of what to them is desirable or undesirable food. Caterers should respond to this demand, rather than imposing what is convenient or dogmatically required of them.

The taste and smell of reheated food

Despite the commonsense view that cooking food twice cannot improve it, and may well cause it to deteriorate, some caterers claim, and also apparently believe, that food is so improved. The problem is that scientific criteria cannot be applied. How do you measure taste or smell as far as appeal is concerned? You cannot. These comments are therefore personal.

My first belief is that some foods withstand (but are not improved by) two phases of cooking better than others. These include pasta, mince and stews, and some types of eggs and carrots. Fish, most green vegetables and potatoes deteriorate most rapidly. This means that cook–freeze and cook–chill operations rely heavily on those foods most amenable to surviving the system and not necessarily those that should be preferred. No wonder the typical hospital meal is stew and carrots!

Reheating is particularly liable to damage the texture of vegetables, and fish and other foods high in polyunsaturated fatty acids generate an acrid or pungent smell because of the decomposition of these acids on storage after first cooking.

With chilled food, the greatest deterioration occurs after 3 days, and it is no wonder that so many meals require spices to mask these unpleasant tastes.

This is no way to produce food, and it is not surprising that where these systems are in use there is increasing dissatisfaction with the final result. There is one large local authority in England that produces 25,000 reheated cook–freeze meals daily for school lunches. It is no surprise that only 13,000 are actually eaten. Many discerning parents provide their children with packed lunches. The problem is that because of the benefit to the catering, and the alleged financial savings (the scale of which must be small; indeed some must be more expensive than conventional), many institutions have invested millions of pounds of capital into producing substandard food. The return to conventional catering can only be expected to be gradual, but the advantages must be obvious to the diner, and the net result in ergonomics will be to find work for many unemployed people, and reduce the need for new equipment and the colossal amount of energy wasted on freezing/chilling and reheating.

But there are optimistic signs. Some New York Hospitals have returned to cook–eat from cook–chill! Perhaps America is going to lead the way in admitting that cook–freeze and cook–chill were terrible mistakes? I hope so.

17
Enjoying eating

Before giving suggestions for eating healthily and for pleasure, I will put most of you into one of three categories, based mainly on differing attitudes to eating meat. It is not the purpose here to recommend or question any one category, but to accept that different individuals do have varying philosophies. This diversity has become accentuated in recent years, particularly in the UK, and no doubt the debate will intensify in North America in the future.

Category 1

You are the traditional omnivore. Meat, and on occasion fish, provides the main centre piece of the meal. Bread, other cereals, pasta, potatoes and vegetables are primarily adornments, although you do appreciate that some vegetables anyway are essential. You look for quality in the meat (and fish), as you find it on the plate. You appreciate tender steak, veal, lean ham, and poultry. You have managed to dissociate the conditions of the rearing and slaughter of the animal from the item on the plate.

In North America, you will not have been troubled by news of the British Mad Cow Disease. In the UK, you will have accepted the reassuring view of the British government, so beef continues to be your main meat.

Category 2

You are still biologically omnivorous in attitude, but you do not give meat as much significance as in Category 1. You will be happy with a pizza with sometimes a token content of meat or none at all. You do not believe that bacon, eggs and sausages are a must for breakfast. You are also increasingly aware of the conditions of intensive meat and fish production, and are troubled by the approach to rearing food mammals and birds in confined and cruel conditions, to increase efficiency of meat production. You will tend to favour free-range or organic meat: lamb, wild birds, and also deep-sea white fish. But many of your meals may not contain any meat or fish at all, and your total consumption of meat has dropped over the last few years.

Category 3

In this category, you are essentially vegetarian. Not that all of you believe that meat is biologically inappropriate for man, the omnivore, but that it is morally indefensible to rear and kill mammals, birds and fish for food. Your numbers have grown in recent years, and you can now point to the increasing acceptance that an exclusively vegetarian diet can be nourishing and enjoyable. You also point out that feeding captive animals with cereal feed is a highly inefficient process, and it would be preferable for us to eat the cereals directly. You eat free-range eggs (and deplore the battery system), and dairy products. Your only difficulty occurs with questions, for example, over the fate of the unwanted male calves incidental to dairy products and the feeding of carnivorous pets, such as cats. Vegans go further than you and find all animal products unacceptable.

Historically, up to the 1930s and 1940s, the great majority of people were necessarily in category 2. There was little intensive rearing, and meat and fish were a respected luxury and eaten in relatively small amounts. The expansion of intensive farming mainly in the 1940s and 1950s created an abundance of meat, and we ate more. The 4 oz. steak increased to 16 oz. (I have seen one menu with 32 oz. rump steaks), it came to be eaten daily, and then for almost every meal. Chicken, once a rare luxury, became a routine provider of cheap protein. Eggs have never been so cheap in absolute values as they are now. Many of today's dietary difficulties are caused by such availability of cheap meat, fish and eggs. The rise of vegetarianism has been one response to this. So let's make some suggestions as to how each of your categories can eat well. At this stage, recipes will not be given; rather a general approach to eating will be suggested. At the end of the chapter, some of my mother's own favourite recipes are listed. I was brought up on them. They are thoroughly recommended!

1. The meat enthusiast

We shall base our eating on three meals a day; in no way need this number or timing be kept rigid. It is the total daily intake of food that matters.

Let's plan the main meal first – say, in the evening – and then work out the other meals. As a meat eater, you want quality before quantity. A 4–6 oz. tenderloin or fillet steak grilled or broiled to your liking should provide most, if not all, of the daily protein you need. Eat more than this, and you are being greedy! There is no better accompaniment than a jacket potato, but most restaurants now ruin them. This is one way really to enjoy one.

First, the variety must be right to produce a light floury texture. It should be quite large and oven-baked (not microwaved), naked, not covered in foil – this means that the skin will brown a bit. On serving it should be cut in half longways and just one half should be enough. Sprinkle a little salt and pepper over it and a knob of butter. The aroma from the partly burnt skin should mingle with that from the steak. There is no reason not to eat the skin – at least most of it.

I've mentioned the dreaded 'B'-word. But I prefer butter to

cream cheese or yogurt on jacket potatoes. So no apologies, although we will have to bear this in mind with the rest of the meal.

With the steak and jacket potato, I suggest a salad. I do *not* mean a heap of chopped stale lettuce leaves, topped with under-ripe, tasteless, fleshy pink tomatoes, and wedges of inedible lemon. Try this. On a separate dish arrange whole young lettuce leaves overlapping on the base. On these arrange slices of peeled orange (against the grain, as it were), add some chopped apple (one of the occasions you will eat an apple) sprinkle with nuts, either pine nuts or grated walnuts, and anoint liberally with dressing of vinegar and sunflower oil, seasoned with basil or mint. Then do something quite unexpected. Eat it all!

For starters and sweets, go for something 'fairly neutral', such as soups (a favourite is fish), prawns, or a few nachos and dips. But there is really no compulsion to add chilli flavour to everything, not even in the USA. For dessert, I suggest a water ice.

This is a fairly well balanced meal, though perhaps a little too positive in saturated fat. Don't worry about this at this stage, as if the rest of your intake is on the light (sorry, I should say 'lite') side, then you will burn off the excess fat.

For breakfast, I would suggest modest portions of two of the following: high bran cereals, muffins, toast from wholemeal bread, pancakes. Eggs and bacon is an unnecessary extravagance.

For lunch, the key is not to eat too much. I'm sure you are already thinking 'burger'. If you are, I suggest either a basic burger with some type of potatoes and salad, or one of the complicated burgers which now seem more akin to laminates of woods with so many layers of mince/cheese/ketchup/onions, etc. within the bun. Don't eat anything else.

But I don't really approve of burgers, as explained in Chapter 13. So how about some instant chicken pieces or a sandwich. The fears over contamination of tuna do seem overdone, but how about a bacon/lettuce/tomato filling?

What to drink? In the USA you find amazingly tempting bargains for the fizzy brown drinks – you know what I mean! There are two problems: the environmental impact of disposal of the cans, with which I hope we are coming to grips (see Chapter 12);

and the sugar. If you drink regular sugar-sweetened cola drinks, you will need to forgo some of the other components of the diet. The choice is yours: either two cans of cola, or that jacket potato! There is nothing actually dangerous, illegal or antisocial in drinking tap water, except that it is not promoted on television, so no-one does it!

With brown fizzy drinks, the message is simple: drink sugar-free, and ensure recycling of the cans. The vogue for caffeine-free derivatives is difficult to uphold to such an extent. Only a tiny minority of people are genuinely intolerant of caffeine. The process of decaffeination must waste energy, increase the cost, and is always liable to lose some of the subtle aroma.

Is the picture beginning to emerge? These are suggestions relevant to both North America and the UK, and apply to eating in and eating out. You could find many alternatives, but the type of food suggested, if eaten in moderation, should be ideal for all ages.

2. The kind omnivore

As before, we will plan our day's food from the main meal and then fill in the gaps. The first realisation is that you have more choice once you abandon the compulsion to base your eating on a huge steak. But keep with beef for the moment; just a couple of ounces of quality minced beef (e.g. sirloin) used with pasta (made with eggs) could provide most of your daily protein. But in this category, you may not feel happy with beef because of the intensive rearing of cattle in North America or the BSE issue in the UK. Sheep are necessarily free-range, traditional joints such as leg or shoulder are justifiably popular. Yes, the meat is often fatty, but there is no need to eat the fat and this is most easily avoided with lamb chops. New Zealand lamb has a lot to recommend it; there would appear little pressure for the use of drugs and intensive rearing in that country. A typical serving of 2–4 oz. may seem a bit meagre, so you might like to add, say, an onion sauce (lightly cooked onion rings in a white or bechamel sauce) or even Yorkshire puddings? These are made of batter (eggs, flour and milk) as for pancakes, but oven-baked, and history tells us that they should be eaten at the start of the meal to reduce your appetite for the tiny meat serving to follow.

Or you could venture to the luxury foods of wild duck or pheasant, quail, or even wood pigeon. Venison comes into this category and needs to be approached with caution. Whilst some deer really are wild, many are reared closely in fields, and some are effectively shed-reared and fed artificially. The problem is for the consumer is how to know, when buying meat or eating it in a restaurant, the exact source of the product.

Then there is fish. There are claims and counterclaims as to whether fish feel pain, but there must be differences in attitude towards say putting live lobsters into boiling water or the trawling and netting of deep-sea fish. (Purely as an aside, I personally do find the former unacceptable but not the latter).

The problem with most of our inshore fish around the UK and in parts of North America and elsewhere is the risk of sewage and industrial contamination, and subsequent diseases such as Hepatitis A. This means that unless we are sure of the source of the prawns, shrimps and so on, there has to be a question mark over these. Not so with most deep-sea fish. The main problem here is overfishing. The solution to the apparent dilemma between an abundance of contaminated shellfish and a scarcity of the safe deep-sea fish is obvious: eat less of the deep-sea variety! Let me explain. With fish, e.g. haddock, plaice or sole, we have become habituated to eating them whole or nearly whole. Rather like steak, the fish should cover much of the plate. Seemingly, when it does there is a move to bigger plates, often oval-shaped to hold even bigger fish. How greedy can we get!

The problem is that many of those whole fish so desirable to the restaurant-goer are immature, and are caught before they have reproduced themselves. It is one of the few occasions that I can compliment the UK Ministry of Agriculture, Fisheries and Food on taking a lead in ensuring that only certain larger sizes of deep-sea fish are caught.

Fish flesh is a very potent source of high quality protein, polyunsaturated fats and vitamins. A 4 oz. helping per day is quite adequate, and incidentally there is nothing reprehensible in eating the skin.

You can now see the answer to the fish problem: that is, to eat sections (or steaks if you like to call them that) of fully grown fish that would at least have had the chance to have bred before capture.

A naked slab of sole or whiting on the plate may not appear that inviting and look on the mean side, hence the need for a sauce. Old-fashioned parsley bechamel sauce simply made with flour, milk, sunflower oil and seasoning, is a suggestion. Incidentally, 'white' sauce made from the brownish wholemeal flour is still surprisingly pale – and, to many people, tastes better.

Recently, I was served grilled sole in a Toronto restaurant with a bright red sauce, heavily seasoned with chilli! The subtle flavour of the fish was completely dominated.

With each of these meats or fish, the amount suggested should provide the necessary protein. Those of us with the habit of eating more protein than we actually need, must convert the excess into carbohydrate used for energy, or into fat. If the amount of meat or fish does decline, then we might need more carbohydrate. Gone are the days when the world had a neurosis about carbohydrates and diabetics. Enjoy your spuds!

In addition to jacket potatoes already mentioned, I would like to commend genuine new potatoes with parsley or mint and a little butter. But they must be fresh and cooked to order. That is the problem.

In mentioning vegetables here, in addition to salad, can I make a plea for the *seasonal vegetable*. I know that many recipe books and restaurant menus require the inevitable availability of any named vegetables, but does not universal availability of any one vegetable reduce its quality at certain times of the year?

Take beans, and a type which I shall refer to as runner. These grow on climbing plants or vines that have orange flowers and are produced in the Northern Hemisphere between June and September. When freshly picked and sliced and cooked for a few minutes in boiling water they are delicious, firm, bright green and full of flavour. The frozen type always seem to lose colour, texture and flavour, and imports can become limp, dull and rubbery.

For breakfast, eggs or bacon are not compulsory. It is true that eggs provide a good source of protein and vitamins, but we probably are eating too many overall; and the egg yolk is abundant in fats, and might have a touch of salmonella in it. One of the best features of North American food is the availability of breakfast muffins – oatmeal, raisin, raisin bran, low fat, bran, lemon, cranberry, blueberry, choc-chip, apple-'n'-spice, carrot, to name a few!

Yogurt is deservedly popular and if you are worried about insufficient protein from your main course, then yogurt may be the answer – or milk, for that matter! Let's assume that your main meal and breakfast are not over-indulgent and that any bread accompanying the meal is wholemeal (with butter) then you should be able to treat yourself to almost anything for lunch up to a limit of 500–600 calories.

3. The vegetarian

Whilst we have considered meat substitutes in Chapter 8, we have not considered the whole daily diet of the vegetarian. It can be nutritious, enjoyable, and economical.

By looking at the ways we can quite easily reduce our meat intake, readers should not find the move towards ridding the diet completely of meat at all surprising. In the UK, the number of people eating no or very little meat or fish is now around 4 million, and I am sure this trend will continue and spread to other countries. In the UK the Vegetarian Society is a highly laudable and credible organisation that works to explain and promote vegetarian food. Its magazine is excellent. The following points are my own and are derived from some of the dishes known to the author or even cooked by him!

But first I want to dispose of the myth that a satisfactory vegetarian diet is more difficult to follow that an omnivorous one. It is true that some care is needed to ensure adequate protein intake, particularly for children and pregnant women, but there is rarely any concern over adequate polyunsaturated fats, vitamins, or fibre. All vegetables do contain certain amounts of protein and with whole cereals, peas and beans that amount is considerable. The one caveat here is that reliance on just one type of cereal or pulse could leave you rather short in one or two of the essential amino acids that make up protein. But this is very easily avoided, by eating various combinations of plant foods, and in practice a vegetarian diet should be preferable nutritionally to meat-based food.

One of my favourite main meal vegetarian dishes I shall describe as bean casserole. It has other more sophisticated names that tend to confuse. In essence, haricot (as in 'baked' beans) or similar beans are combined with sauces, onions and other vegetables, and partly cooked before being put into individual or 'family' casserole dishes. The top is given a defined structure with, say, sliced tomatoes, or courgettes, flour and grated cheese. The baking time will depend on the thoroughness of previous cooking. Experiment for yourself. As accompaniment, use seasonal vegetables and salads as already described.

Alternatively, use the bean casserole like a hungarian goulash within a circular boundary of rice. I defy anyone to claim that a vegetarian goulash is less appealing to the eye than the meat goulash.

Alternatively, a main item of cauliflour cheese or cheesy potatoes could be considered. Four ounces of hard cheese provides a great deal of daily protein, and the potatoes can be delicious. Try this. You will need a large flattish casserole dish filled with layers of thinly sliced potatoes, sprinklings of flour and part-cooked onions. After adding milk to around half full, cover the top thickly with grated cheese then a little flour and bread-crumbs. Then bake for about one hour at 375 °F. Delicious. Nutritious.

If there are concerns about adequate protein in the diet, you can make this up with breakfast with eggs (well-cooked, of course) and yogurt, and at all times enjoy quality fresh wholemeal bread.

For lunch you may want something instant. Pizzas are made for you, but the quality is so variable and even a few minutes' delay on travelling can wreak havoc. The crisp edge goes hard, the firm centre goes limp and soggy and the topping becomes sticky and lifeless. The traditional and simple old-fashioned pizza with tomatoes (Italian of course), mozzarella cheese, perhaps livened up with olives, capers and other herbs is a favourite of non-vegetarians!

You will be familiar with the various alternatives – quiches, meat-free and fish-free sandwiches. The word sandwich these days includes a number of products where a filling is enclosed between two parts of roll, bap, bun or bread and as such can be eaten in the fingers. They can also be made in advance of sale, and the quality deteriorates with storage, with the bread becoming soggy or the filling dry. Meat and fish-based tend to last relatively well so are popular with the manufacturers and consumers. The key to the meat-free sandwich is its freshness. My favourite is made from that day's wholemeal bread, a wipe of quality margarine, with slices of cheese (not grated) and pickle.

Snacks

Hot dogs

Have I the nerve to criticise America's No. 1 snack food in a book that hopes to sell in the USA? Yes, I have! Take a typical hot dog bar. You see the lines of pinkish sausages cooking on their rollers, with fat trickling over them. The smell is unmistakable and some of us have learnt to like it, if not actually *want* it. At the queue, expectation mounts. The bread rolls are displayed there too, nicely browned on the outside. The promotional picture shows a man giving into temptation with a fine line of mustard zig-zagged along the sausage.

The placard says '100% pure beef'. The expectation is reinforced. Beef is good for you and if it is both 100% and it is pure, it must be wonderful (don't ask me what 100% unpure beef might mean, because I don't know).

The first disillusionment comes after paying. You are handed a hot foil package which, on opening reveals a long pink object, the texture of brittle rubber. You gulp it down quickly, not so much because you do not wish to savour the pleasure, but because you have succumbed to the selling propaganda and bought two for nearly the price of one. Still, hot dogs symbolise American manhood and two must be better than one.

But then you start thinking – or rather, you are now – and wonder how 100% beef becomes converted to granular rubber. Does 100% beef always mean all steak: that is, the animal's muscle? No, it does not. Countries have their own definitions of how words can be used to promote food, but in general 'beef' means the edible part of male cattle and indeed also the poor old cow at the end of her few years of milk production. Beef might mean high quality steak, but it can also mean the residuum of the animal recovered by mechanical scrapers and all the other parts we don't like to think about. There must be substantial fat present to gel the sausage together and also to cook it.

I don't like hot dogs, in the same way I don't like burgers.

Instant chicken

There has been, certainly in Europe, a trend towards eating chicken meat rather than red animal produce, on the assumption that white meat is healthier. Specifically, chicken is thought to contain less saturated fat. This may not actually be true, unless one carefully discards the chicken skin and underlying fat, the main source of flavour. By the time that chicken meat has been processed into lumps or purée, or sliced and incorporated into burgers or sandwiches, has it any positive taste at all? No. Is that why it is popular? No taste = no complaint? No wonder so much chicken has to be encased with flavoured coatings!

I don't feel particularly antagonistic (believe it or not) towards instant chicken lumps, legs, wings and so on. But the popularity of otherwise inedible wings, spiced with the inevitable chilli sauce and called buffalo wings, must be a reflection on the total absence of natural chicken flavour. Is it really justified to raise broilers under intensive cruel conditions just to provide protein?

Why not coat soya protein with exciting flavours? At least with these products, there should not be a problem with salmonella, campylobacter and the rest. If you take the positive marketing image away from chicken, and think how broilers are produced, slaughtered and processed, can you justify these products at all?

Having established that the average diet, whatever your philosophy over meat, is likely to contain an abundance of protein, instant chicken immediately becomes capable of being discarded.

Popcorn

I like popcorn. It *is* exploded seeds of maize and I don't think it can be faked. It is also fairly filling. If there is a problem it is the substances that might be added to excess, such as butter or sweetening. But as a snack, OK.

Chips (Crisps), biscuits etc.

We have already mentioned these in Chapter 8, but considering them as part of the daily diet I want to reinforce the point that despite their similar presentation and appearance, crisps or chips vary enormously.

As provider of energy and to fend off the pangs of hunger, there is nothing wrong with the composition of chips. It is the misleading advertising slogans that concern me.

In response (or do I cynically suspect as the cause of) the massive cholesterol neurosis sweeping North America, many crisps are claimed to be 'cholesterol free'. The idea presumably is that high amounts of cholesterol in the blood *cause* heart attacks (this actually has not been been adequately researched or proved: see Chapter 15) and that if you don't eat cholesterol it cannot clog up your arteries. But crisps, unless dipped in egg yolk or animal lard, should not contain significant amounts of cholesterol anyway.

This positive exercise in misinformation is further exacerbated by the failure to make prominent on many brands that they have been fried in hydrogenated vegetable oils. These are unnatural and may well be dangerous, and it is hoped that readers demand that the regulatory authorities print a warning on the packet to possible dangers (that is in putting your blood cholesterol up as a result of their effect in the body after eating) and that they assess hydrogenated oils in the same way as they do other food additives. Will the American Food and Drugs Administration please act? The Dutch government has already acted over margarine containing some hydrogenated oils.

Those crisps (or chips) which are recommended are those that contain potatoes and pure vegetable oils, such as sunflower, canola (rape seed), and cottonseed, which are high in polyunsaturates and monounsaturates. Palm oil is, of course, a vegetable oil and is saturated, and is presumably no better for you than beef fat.

The low cholesterol and cholesterol-free misinformation extends to all manner of processed food, such as waffles. I wonder how many people fervently select cholesterol-free waffles from supermarkets but unquestioningly eat waffles from restaurants. Indeed, how many people ever think of questioning the source of food in restaurants? No, the décor of the room, the colour of the waitress's hair, and tablecloth are all important. If you are worried about eating too much cholesterol (and most people should not be unless you have a serious medical condition: see Chapter 15) the approach is simple – avoid egg yolks, and do not be too lavish with liver, kidney or shellfish.

Ice-creams, frozen yogurt

Some of the cholesterol phobia has spilled into these products, and very few of these have significant amounts of cholesterol, even those made from animal fat. But the chief development in this area has been towards frozen yogurt, particularly low-fat types. This is desirable on the grounds of reduced calories alone, but the yogurt does also contain more protein than traditional ice creams. As already stated, very few people have a basic diet defective in protein so this is hardly relevant. But frozen yogurt is recommended and I do not believe that it is harmful to take snacks between meals as long as the total daily intake is not excessive.

On a recent visit to the USA I enjoyed a delicious blueberry frozen yogurt, although its enormous size surprised me.

Cholesterol in most snacks is a giant red herring.

My mother's 14 recommended recipes

LENTIL SOUP

Ingredients

½ pint red lentils
1 medium onion
1 medium carrot
1 oz. butter
pepper and mixed herbs to taste
2 pints stock in which ham of hock has been well cooked after washing

Method

Wash lentils and strain, then soak in some of the ham stock, according to instructions. Melt the butter in a saucepan, add the chopped onion and carrot. Cook gently for 10 minutes. Add pepper and herbs and the stock. Bring to the boil and add the lentils with the stock in which they have been soaked. Cook gently until lentils and vegetables are tender. Add the best of the ham from the hock, cut into small pieces. Add salt to taste only if required.

WHITE VEGETABLE SOUP

Ingredients

½ oz. butter
½ lb. prepared white vegetables i.e. leeks, turnips, celery, fennel and potatoes – as available
1½ pints good light coloured stock, e.g. chicken
1 level tablespoon flour
¼ pint milk
1 tablespoon cream
white pepper, salt and teaspoon of sugar

Method

Cut prepared vegetables into strips and toss in the melted butter in a saucepan for 5 minutes; do not brown. Add the boiling stock and seasoning and boil gently until the vegetables are tender. Mix the flour with the milk until smooth, add to the saucepan and cook for 5 minutes. Remove from heat and add the cream. Serve at once with sprinkling of freshly chopped parsley or sorrel.

FISH AU GRATIN WITH EGGS

Ingredients

1 lb. white fish
3 eggs, hard-boiled and shelled
½ pint white sauce
2 oz. mature Cheddar cheese, grated
½ teaspoon mustard powder

Method

Grease an oven dish. Skin and bone the fish and place in dish.
Place eggs cut in half between the fish. Make the seasoned white
sauce, adding the mustard and grated cheese. Pour over the fish
and eggs. Scatter a few white breadcrumbs over the top. Bake at
190 °C (375 °F) for 30 minutes.

FISH PIE

Ingredients

1 lb. cooked fish
½ pint parsley sauce
1 lb. mashed potatoes
1 oz. butter
pepper and salt
browned breadcrumbs

Method

Remove bones and skin from fish and flake it. Fold the fish into
the sauce and place in a greased pie dish. Cover lightly with the
mashed potatoes, then the butter cut into small flakes. Sprinkle
the browned crumbs over and bake at 200 °C (400 °F) for 20
minutes.

LANCASHIRE HOT POT

Ingredients

1½ lb. best end of neck lean lamb chops, separated
seasoned flour
2 medium onions, peeled and sliced
2 lambs kidneys, cored and diced

1½ lb. potatoes, peeled and fairly thinly sliced
¾ pint stock
a little lard
(Optional – 1 oz. butter)

Method

Trim excess fat from chops and coat in the seasoned flour. Melt the lard in ovenproof casserole on hob heat. Fry the chops to brown on either side. Then arrange a layer of the onion and the potatoes. Repeat this, finishing with a layer of potatoes. Put small dots of the butter over the potatoes. Pour the stock gently all over. Cover with lid and cook for 1½–2 hours in a moderate oven, 180 °C (350 °F). Remove lid and continue to cook for ½ hour high in the oven to brown the potatoes.

This dish can be made the previous day and reheated.

SWEET AND SOUR PORK
(good served with boiled or fried rice)

Ingredients

1½ lb. lean pork
1 egg
2 tablespoons flour
sweet and sour sauce
2 tablespoons of milk
pinch of salt
oil for frying

Method

Cut the pork into 1 inch cubes and dip into the batter made from the flour, milk and egg. Cook in the hot oil until the meat is tender and cooked well and the batter is crisp and browned – about 10–12 minutes. Drain on kitchen paper.

The sauce

2 oz. sugar
1 teaspoon cornflour
1 tablespoon soya sauce
3 tablespoons malt vinegar
¼ pint water

Method

Blend all in liquidiser for ½ minute. Cook in small saucepan until it thickens, stirring all the time. Add 1 tablespoon of one or more of the following: sweet mixed pickle, chopped gherkins, chopped olives, chopped pickled onions, chopped dessert apple or chopped cucumber. Pour over the pork and serve.

CRISP PORK CHOPS

Ingredients

4 pork chops
1 small egg
salt and pepper
3 oz. white breadcrumbs
1 small onion, finely chopped
parsley, finely chopped
lard or oil for frying

Method

Mix together all the ingredients, apart from the chops, and place on a plate. Rest the chops in this mixture for half an hour, turning once or twice, and pressing well in. Fry in the hot fat, turning once until chops are well cooked through, reducing the heat if necessary, according to the thickness of the meat.

CHEESE PUDDING

Ingredients

6 slices thin bread and butter
6 oz. mature Cheddar cheese, grated
3 medium size eggs
1 pint milk
salt and pepper

Method

Butter a pie dish. Arrange in it the bread and butter in alternate layers, with the grated cheese, topping it with the bread and butter. Beat the eggs well and add to the milk and seasoning. Pour over the contents of the dish. Allow to stand for about 10 minutes before baking at 180 °C (350 °F) for 45 minutes or until set.

CAULIFLOWER AU GRATIN

Ingredients

1 medium sized cauliflower
4 oz. grated cheese – hard mature Cheddar or Parmesan
white sauce made with ½ pint milk, 1 oz. flour, 1 oz. butter, white pepper and salt

Method

Wash the cauliflower and trim stalk and outer leaves. Place stalk upwards in boiling salted water. Cook until the flower is just tender. Strain, and put flower upwards in oven dish. Pour over the white sauce, having added 3 oz. of the cheese. Sprinkle remaining cheese on the top, and bake near the top of the oven for about 10 minutes at 220 °C (430 °F).

CHEESE STRAWS

Ingredients (makes 50–60 straws)

4 oz. plain flour
3 oz. finely grated hard well flavoured cheese
2 oz. butter
1 egg yolk
pepper, salt, and if liked a tiny amount of cayenne

Method

Mix the flour, cheese and seasoning. Rub in the butter. Mix to
a stiff paste with the egg yolk: you may not need it all. Knead
very well until smooth. Roll out on floured board to ¼ inch thickness. Trim into strips 2½ inches wide. Cut these strips very evenly
into straws about ¼ inch wide. Re-roll trimmings and repeat until
all the pastry is used. Place carefully, slightly apart, on lightly
greased baking trays, probably needing three, and bake for 10–
15 minutes at 180 °C (350 °F) until pale golden brown. Remove
carefully to cooling rack and pack into dry airtight tin as soon as
cool.

CHEESE SOUFFLÉ

Ingredients

2 oz. butter
2 oz. plain flour
½ pint milk
3 large eggs, white and yolks separated
4 oz. finely grated hard Cheddar cheese
salt and touch of cayenne pepper

Method

Melt the butter in a medium-sized saucepan and add the flour. Cook for a minute, remove from heat and add the milk gradually, stirring all the time. Return to the heat and bring to the boil, stirring continuously. Remove from heat to cool a little. Add the egg yolks one at a time, mixing each well in, also the seasoning and finally the grated cheese. Whisk the egg whites until stiff, with a tiny pinch of salt. Stir in one tablespoon first and then lightly fold in all the remaining egg white, until fully blended. Pour mixture into 7–8 inch ungreased soufflé dish. Bake in oven preheated to 205 °C (400 °F) for 25–35 minutes, until well risen, golden brown and set all through when tested with thin skewers. Serve at once.

APPLE CHARLOTTE

Ingredients

1½ lb. cooking apples (Bramley's)
4 oz. sugar
a little water
bread – not too fresh
butter

Method

Peel, core and slice the apples – not too thinly – and cover with water in a basin. Cut thin slices of generously buttered bread, removing crust edges. Cut into squares (about 1½ inches). Butter an oven dish and fill in layers with the apples sprinkled with sugar and the bread and butter. Finish with a neat layer of bread on the top. If necessary add more sugar and some water. Bake at 190 °C (370 °F) for about 30–35 minutes, until the top is crisp and golden and the apple tender. Sprinkle top with caster sugar and serve. Improved with a serving of cream.

COFFEE MOUSSE

Ingredients

can evaporated milk (about 12 oz)
1 envelope gelatine
4 heaped tablespoons sugar
½ pint milk
1 slightly heaped tablespoon custard powder
1 level teaspoon powdered instant coffee

Praline

6 oz. white cane sugar. ½ pint of water. 4 oz. almonds, blanched, chopped and roasted.
Butter a 7 inch sandwich tin and put in the prepared nuts, spread out. Make the syrup with the sugar and water in a small pan, without stirring, until pale golden in colour. Pour this over the nuts and allow to harden, before turning on to a board and crunching with a rolling pin.

The Mousse

Whip tinned milk until stiff. Make a custard with the milk, custard powder, coffee and sugar. Cool slightly. Dissolve the gelatine in 2 tablespoons of hot water. Fold all these things lightly together, mixing thoroughly. Turn into a serving bowl and put into the refrigerator to set.

To Serve

Spread a good layer of whipped cream evenly over the top of the mousse. Sprinkle liberally with the praline. Surplus praline will keep well in an airtight tin.

LEMONADE

Ingredients

6 lemons (4 if large and juicy)
3 lb. granulated sugar
2 oz. citric acid (from the chemist)
2 pints boiling water

Method

Peel lemons thinly. Add juice and peel to sugar and citric acid. Pour the boiling water over. Strain when cool. Dilute to taste when using.
Half quantities can be made. This lemonade will keep well in a stoppered bottle in the refrigerator.

PART VI

Is there a solution?

18
Is there a solution?

First let us summarise the problems. Of course, many people are doing well out of modern methods of food production and do not believe that there is a problem at all. But I would not have written this book unless I thought there really was something seriously amiss with the provision of our food.

With development of farming, some deforestation of the world can be justified for the growth of crops, but surely not for the extent of animal rearing, be it free-range or intensive, that has occurred. The sheer inefficiency of feeding cereals to mammals and birds, rather than eating them directly, must surely be undesirable? This must result in more deforestation than is actually needed.

Moreover, intensive agriculture has seriously damaged our soil, and soil deficiencies in the future must include minerals such as phosphates and potash, humus, and of increasing importance, water. One point about global warming that has not been widely appreciated is that as the temperature of the atmosphere rises, it can hold more water as invisible vapour, with less condensed water as clouds or rain. Global warming could generate its own momentum, with the clearer atmosphere allowing greater penetration of the sun's irradiation to the land, and even greater warming. This is the reverse of the transition into each of the Ice Ages, when global cooling would have resulted in greater condensation of water in the air, and hence a progression to even more cooling. The extent of this warming momentum will remain speculative for decades.

Intensively reared animals are associated with other problems:

BSE in the UK; salmonella and campylobacter amongst infections; poor quality meat low in polyunsaturates; the need for drugs; animal cruelty; and the practical difficulty of disposal of animal effluent. Intensive farming of fish, notably salmon, produces inferior quality food relative to salmon in the wild and is damaging to the environment, including the probability of an adverse effect on the vigour of wild salmon. Let the problems of fish farming and BSE in the UK be a powerful warning to our friends in America.

We are wasting the world's finite reserves of energy in many more ways. The concept of cooking food twice as a routine seems very silly from an energy usage point of view. Similarly, the need for instantly available hot meals or snacks is wasteful of energy. The amount of packaging has come under severe and justified criticism over the last few years, and whilst biodegradable or recyclable products must be preferable to polystyrene burger boxes, even more desirable would be a reduction in the amount of packaging altogether, so that then there would be no need to recycle so much material.

The way we now produce and eat food is very recent. In the USA the expansion of highways and the essential role of the motor-car were established in the 1930s, as was the role of eating out. In Europe, most of these changes were not established until after the last World War. I can recall well rural life some 40 years ago, partly through contact with my grandparents.

It must have been around 1950 that I last saw my grandfather. Because of difficulty with hearing, his ear was close to an old battery radio that was broadcasting news of the Korean war. The radio was the only 'modern' device in the cottage that was in deepest East Anglia in the UK. Water was collected in a tank from rain falling on the roof. The lighting was an oil lamp, with candles taken upstairs when required. The toilet was a bucket in a shed in the garden. The heating/cooking was an iron Victorian range in the one living room. The order for non-perishable groceries was taken during one week and delivered the next. Milk arrived from the farm opposite in churns. Bread, meat and fish were delivered each week.

Outside, the front garden of the cottage was sufficiently picturesque to have been the subject of a Helen Allingham painting. To the right of the house was an extensive orchard of plums, apples and pears. To the rear, there was a large vegetable garden, and to the left more fruit trees, sheltering free-range chickens that used sheds for perching and laying eggs.

Winters must have been grim, but I can remember with nostalgia the long summer evenings. There was no television. We talked and played cards or read, either by daylight or from the luminescent mantle of the oil lamp. Everyone sat round a large table, their faces glowing from the lamp, producing shadows on the walls.

By chance I was recently in the area passing *en route* to give a talk. Curiosity got the better of me. I had a look. The first thing I noticed was that the name had changed from Bramble Cottage to Bramble Farm. The building seemed much smaller than in my memory, but all main services were now connected, with an enormous oil tank in the middle of what had been the front garden. The house had been given a face lift, but then partly shuttered up. It was deserted, seemingly now a weekend retreat for Londoners.

The orchard and hedges had disappeared, now incorporated into a great field. The land that had once grown vegetables and reared chickens was littered with rusting derelict vehicles. The flower garden was poorly kept grass. It was just possible to see signs of roses attempting to struggle upwards. (My parents tell me these had been a gift in the 1940s.)

In the village nearby, the shops had all gone, the public house had been demolished and the churchyard was overgrown. It is true that this area had been particularly backward in acquiring the services we now take for granted, but the changes in the way we live in the last few decades are remarkable. But not all the changes are for the better.

It is as if we have not paused in the last 40 years really to question whether these changes we now take for granted are always 'progress'. I don't think that the rate of this change, or indeed the scale of this change, can be supported in the future, without accelerating the impending world environmental catastrophe.

The changes in Europe and USA have been driven by consumer demand, and whether we like it or not, the whole concept of the future must be provided by consumer attitudes. It is for this reason that for any of the suggestions here to be adopted, it must be because you, the consumer, want them to be. They cannot and will not be imposed by bureaucratic dogma, whether from Brussels, Westminster or Washington. To put our provision of food right, which I believe is essentially to revert to the mixed farm, the re-establishment of local retailers and the preparation of most meals in the home, will require major changes in society. These may well be forced upon us when the oil runs out. Far better to anticipate the crisis and act now.

Hopefully, readers will find some of these issues of concern and be determined to see changes. But the author is pragmatic and is fully aware that one book will not change the world.

We can approach the solutions in two ways: to try and identify what our ultimate goals are, and then to see how we can achieve them. This may not be at all easy given the structure of our society.

Cleaning up poultry and eggs

To rid chickens, ducks and turkeys of salmonella and campylobacter must be attempted, whether or not we move to free-range husbandry. The knowledge is available to do this but because these bacteria are so well adapted to growing in poultry, we cannot expect absolutely salmonella-free birds. This, as the producers claim, does not mean that it is futile even to attempt to improve matters. We can and should, and it need not be too expensive.

The elimination of salmonella from eggs should take priority, as it is impossible now for a runny yolk to be reliably safe. But proper cooking of poultry meat should be able to ensure safety.

With eggs, the transovarian means of contamination results in the salmonella passing from chicken to egg to chicken to egg, and so on. The ultimate source of the infection is likely to be in the breeding flocks, although further contamination could occur at other points in the production line.

A salmonella-free breeding flock could be established, with care taken to exclude the introduction of salmonella from the feed or the environment. Blood tests are available to identify contaminated (but not obviously ill) chickens, and a vaccine against the salmonella is also available.

One of the weaknesses in the egg production system is the hatching of fertile eggs in incubators. Here the emerging chicks are not exposed to their natural mother's bacteria, and these are therefore not available to compete with any salmonella that might be present in the chicken. The salmonella can then multiply and produce an illness in the chick. It may well recover, only to carry the salmonella in its internal organs, which can then find its way into eggs when laying starts.

Research has shown that if day-old chicks are fed safe normal chicken bacteria, they are less likely to suffer from salmonella problems.

The approach of the UK government has obviously failed to control salmonella; in essence it has relied on the slaughter of heavily contaminated laying flocks, only to replace them with further contaminated flocks. There has been no real determination to establish a salmonella-free breeding flock.

With broilers for chicken meat, and other birds reared for meat, similar approaches could be used to minimise the risk of salmonella and also campylobacter. One problem here is that even if only an occasional bird is carrying the bacteria in its intestines at the time of slaughter, because of the automated methods of slaughter, plucking and preparation, these bacteria tend then to be widely dispersed. This danger could be reduced, but a small increase in cost would result.

BSE

Mad Cow Disease hangs like a satanic cloud over our British farming, even though the government and its veterinary advisers exhibit an outwardly reassuring attitude. Many aspects of BSE are still uncertain, but one thing is not: that the same type of infection causes BSE or scrapie or the human disease Creutz-feldt–Jakob disease, and that it is so tough it is likely to survive treatment in the rendering plants. Rendering plants are used widely in the developed world, including the USA.

This means that the unwanted remains of animals and birds (including some domestic pets, and I am sorry to have to say so) will contaminate animal feed if they are infected with these agents.

The whole principle of recycling animal remains back to the same, or similar, species must stop – BSE has not extended on a major scale to countries other than Britain, but surely it is only a matter of time, unless action is taken. The system of feeding an animal with remains of its own species is effectively cannibal-ism, and it is now fully established that the brain disease, kuru, in a primitive human tribe of New Guinea (the Fore tribe) was due to just this.

Once an infection of this sort is introduced into a herd or flock of animals, the effort needed for its elimination is very substantial. The first problem is that evidence of this disease manifests itself only several years after infection, during which time the disease could have been silently spreading within the animals.

There is also good evidence from sheep that these types of diseases can be inherited vertically – like salmonella in poultry – meaning that an infected mother (who may appear to be healthy) can pass the infection to her offspring when in the womb. The likely cause of this is the presence of the infection in the mother's blood. If this is the case, the idea that meat (that must contain blood) from infected animals is safe to eat, as suggested by the UK government, is nonsense.

The only certain way of eliminating BSE or scrapie is to slaugh-ter the whole infected herd or flock, because it is not known which animals are infected and which are not, as there is no appropriate test that can be performed on live animals. New animals from a source known not to be infected should be established on new territory. This is required because of the ability of the infection to persist on the ground for years.

This is clearly a massive and costly undertaking and, seemingly, beyond the grasp of the British government. Prevention of such a disease is surely preferable, and this is one very good reason to abandon current methods of animal farming and feeding. How would the USA react to the sudden and widespread appearance of BSE in cattle between Michigan and Texas, and California and Massachusetts? I wonder.

The obsession with efficiency

The separation of the growing of crops from animal husbandry, in the obsession for greater efficiency for each type of product, must be reversed. In the UK at present, because of the folly of this policy, we now have the pathetic and contrived policy of 'set-aside'. Farmers are paid up to £80 per acre to keep land fallow except for two cuts of the grass per year. Whilst this may increase soil fertility, it is not part of a proper crop rotation or a move to mixed farming. The chickens, turkeys and pigs are still housed cruelly in sheds. If the determination of the public to eat cheap chicken and eggs continues, it will ensure that this nonsense is sustained. The realisation that we do not need to eat meat daily, or even weekly, or even at all, still has to come about for most consumers. We should view the chief role of animals and birds controlled in mixed farms as the maintenance of soil fertility. Perhaps we should feel somewhat guilty when we eat them? One simple message for American guzzlers of giant steaks or monster burgers is to eat fewer of them. If you do just this, I'll be happy

There will continue to be undoubtedly ample available land in Europe and North America for the establishment of mixed farms and smallholdings. In the UK, the full impact of BSE cannot yet be quantified but at the time of writing, with the disease showing no signs of abating, a great deal of potential farming land may become available – assuming that we can be persuaded that the infectious agent on the ground will not prove to be hazardous. (We are in the era of uncertainty, but species such as chickens or ducks are most unlikely to be affected, nor will cereals.)

Let us suppose farming methods do improve towards organic, free-range husbandry and the mixed farm, one inevitable consequence is that the price of the product will increase, but perhaps by not too much. This has to be accepted. Some environmentalists have tried to believe that greater costs are not necessarily associated with an improved ecology. But with food, our whole strategy for several decades in the developed world, in particular in the USA, has been towards mass production of cheaper food regardless of the consequences. The reversal of this ethos must increase the price, unless we can keep repeating the miracle of the feeding of the five thousand (incidentally, bread and fish make an excellent diet, as the Portuguese know!). More expensive meat, and probably fish, does not, however, mean that the total food bill will increase, because the alternative vegetable-based food that we will eat could well be cheaper, even than the cheapest burger.

But will the consumer be prepared to pay more for a disfigured fruit or vegetable, in the hope that the condition of the soil will be improved 5 years hence, or that fewer chemicals may be needed? I wonder. Can we really be environmentally committed in a democracy that rates profit as the only criterion of success? I used to believe that this was a dilemma unique to the USA. This is not so, for the British, still very much influenced by Margaret, are now conditioned to the primacy of profit, even though we apparently can manufacture very few goods for ourselves now.

This is the nub of the problem: the theory of what we should be doing may not be capable of being realised in practice – suppose farmers did go 'green', and provide us with environmentally friendly, safe and humanely reared food? Would you buy it, if it was, say, 50% more expensive; or 20% or 10%; or 5% or 1%?

Would not the supermarkets, on your behalf, using their enormous buying power, obtain cheap food from overseas? Have we the will to put farming on a secure ecological footing? Or are we going to continue to surrender to market forces, with even cheaper food demanded by an expanding world population, with the inevitable result in environmental damage sooner or later?

These arguments are not novel, and the author is not claiming to be 're-inventing the wheel', but they do show how difficult our problems will be to solve.

In a similar way, the concept of reversing our expectation that motor-car ownership is essential to family and business life is exceedingly difficult. Looked at in terms of energy and pollution, the day of the family driving off to an instant restaurant for each of its meals is quite unsustainable. Is this really man's biological role? After all man is thought to be a social animal, yet from the way that we eat and behave, we are becoming increasingly antisocial. From the drive-in burger delights, to the individual microwaved ready meal, to the ambulant diner and his bag of fries and lumps of chicken, we are moving away from the meal as a dedicated family event. What we seem to want is instantly available hot meals, and it is no wonder food poisoning is so prevalent.

I used to idolise the French way of eating, and their elevation of eating to something more important than that of solely a body function. I still do to an extent, and locally produced French food can be quite delicious. But the French are succumbing to packets of convenience meals with lurid pictures as to how food used to be, and are also entering the fantasy world of the food image makers and the giant out-of-town hypermarkets.

The choice is yours

We saw, in Chapter 15, that the ideal diet is much more flexible for most people than you probably thought. But the chief uncertainty is the exact composition of any product's nutrients. Wild salmon *is* better than farmed salmon, fresh vegetables *are* preferable to reheated vegetables. The simple description of the food is not adequate to identify its nutritional content. Additive-rich processed food is undesirable not so much because of the uncertainties of the effects of the additives on you, but because the processing must cause decomposition of useful nutrients. The concept of favouring fresh whole food produced locally should not come as a revolutionary shock. It is common sense. But we do seem to want all our food types to be available all the year round. No wonder the quality deteriorates. It always amazes me to see shoppers buying cans of peas in midsummer when pods of fresh peas are available. Yes, I know there is the trouble of actually shelling them, but surely the effort is worth while?

Over the last 100 years, the population movement in most developed countries has been away from the country into the cities. This has accelerated recently and the food industry has responded to it by making more food available in the cities. With some items, staleness must increase, particularly with imported products and their long, tortuous journey to the consumer. It is not the purpose here to denigrate living in cities, but some problems in obtaining fresh food must ensue. This situation accentuates the paradox where much unemployment and poverty abounds in the cities, yet the depopulated countryside is simply crying out for some tender loving care.

Perhaps much of our modern attitude to eating is governed by the ethos of the urban dweller, frequently isolated, or even solitary, impatient and pressurised; *Homo sapiens urbanis* (or *Homo supermarketens*) has succeeded in effecting near complete mental disassociation between what he eats and the conditions of the production of the plateful. Food has become a disposable commodity like newspapers, toilet paper or toothpaste. True, the wealthier members have their paraphernalia of modern technology such as video recorders, fleets of cars, collections of televisions and swimming pools. The food this species eats is subservient to these, although perhaps the tumbler of ice cubes in waiting for the liquor is not. Poorer members of *Homo sapiens urbanis* have few material possessions: much of their meagre income is wasted on unnecessarily complex processed and packaged food. They are fodder to be exploited by the makers of junk food.

No doubt urban man sees himself as the glory of Planet Earth, but he can be sustained in the long term only if rural man is allowed to put his house in order, and the wasteful and environmentally damaging effects of much of our food production are addressed. The most potent force that will continue to counter any attempts at ethical means of food production is the growing world population. Procedures to address the crisis are beyond the scope of this book, but addressed it must be. The tragedy that the momentum towards increased efficiency in agriculture has been associated with increased world population, including the numbers of the malnourished minority, is now painfully obvious to all. Once basic food commodities such as rice, wheat or beef became an international commodity we can see, with hindsight, that this became inevitable. For our methods of food production to become socially and ecologically acceptable, we must address the problems of the excessive numbers of *Homo sapiens*.

Glossary

AD LIBITUM The making available of a constant supply of feed for mammals or birds. Usually stored in high chambers and delivered by gravity to feeding troughs. This enables the mxiumum rate of weight gain of animals and birds under intensive conditions.

ALBINO Living creatures that do not possess the appropriate genes (i.e. DNA) to make pigment, and so look white, pale or colourless.

ANTIBIOTIC A substance produced by one tiny organism, usually a bacterium or a fungus, that can be extracted and purified to be used medicinally in the treatment of infected patients. Some of these natural compounds can be improved by synthetic processes, and for reasons not fully understood, animals fed these drugs increase their weight gain.

BACTERIUM A tiny organism, invisible to the naked eye, so small that about ten million could be accommodated to the head of a pin. Some bacteria grow and divide very quickly, which is a serious problem if they find their way onto food kept warm. The correct way of describing any single type of bacterium is using two italicised words. The first word begins with a capital letter and refers to the genus or general group, and the second letter refers to the specific type. Thus *Staphylococcus aureus* is found in the human nose but when this and related bacteria are described collectively, they are expressed simply as 'staphylococci'.

BACTERIAL TOXIN A protein substance released by some growing bacteria in food, water, or in the human body. The toxin produces damage by interacting with one or more tissues. Some toxins are resistant to cooking temperatures; others are sensitive.

BROWN FAT Brown-coloured tissues appearing fatty but with the remarkable ability to convert nutrients – particularly when in excess of need – to heat that is then lost from the body. People well endowed with brown fat may be able to eat amounts of food usually thought excessive without putting on weight.

CAMPYLOBACTER Bacteria found in the intestines of many birds and mammals, including poultry and cattle. Whilst not commonly causing disease in these hosts, these bacteria can cause colicky abdominal pain and diarrhoea in people. The disease is most prevalent in June, for reasons not understood. Mainly spread by unpasteurised milk, poultry and perhaps water.

CARBOHYDRATES Large natural chemicals made of carbon, hydrogen and oxygen. The proportion of hydrogen to oxygen is always 2:1. Examples include sugar and starch.

CHOROFLUOROCARBONS (CFCS) Synthetic chemicals that can exist as either a fluid or a gas, used as coolants in refrigerators and deep-freezers, and also propellants for some aerosols. Escape of CFCs into the high atmosphere has been thought to have chemically reacted with, and removed, ozone (chemically O_3). Ozone is responsible for shielding people from certain dangerous types of irradiation from the sun. One consequence of this is a predicted increase in cancer of the skin over the next 40 years. There is an urgent need to cut down the use of CFCs and many responsible developed countries have future targets to control their use.

COOK-CHILL A catering system whereby whole meals or their component items are cooked to an initial temperature of around 70 °C, rapidly chilled to just above freezing temperatures, preferably between 0 and +3 °C, and stored and transported prior to reheating and being eaten.

COOK-FREEZE Similar to cook–chill except that the food is chilled down to and stored frozen between −18 and −23 °C.

DDT An insecticide chemical known as a chlorinated hydrocarbon. Whilst effective, its use was associated with problems of extreme persistence, resulting in severe environmental damage, notably to birds.

FOOD POISONING In the strict sense, it is the cause of illness that results from the presence of pre-formed toxins in the food in question. Now also applies to problems of food contaminated with bacteria or viruses in addition to toxins.

FUNGUS A microorganism usually endowed with an invisible (or nearly so) network of root-like threads and large fruiting body (as in a mushroom). Yeasts are also a type of fungus but do not contain the branching threads.

GENETIC DIVERSITY The range of different types of functioning DNA in a species.

GENETIC ENGINEERING The creation of new life forms by reassembling the DNA of different members of the same or even distinct species.

GREENHOUSE EFFECT The postulated cause of the recent and predicted rise in average temperature throughout the world. This is so-called because in a greenhouse, the temperatures rise on sunny day because of the building's structure, which prevents the release of generated heat into the atmosphere. The global greenhouse effect results from heat generated on or near the surface of the earth failing to escape as a result of increasing amounts of carbon dioxide (from burning of oil, coal and gas) and methane (from cattle).

HELICOBACTER PYLORI A newly recognised bacterium, that causes indigestion and probably also stomach ulcers.

HORMORNE Natural hormones are chemical substances produced by one organ in the animal that control the activity of others. For example, the hormone production from a small area underneath the brain induces the onset of lactation after childbirth. Artificial hormones may be exact copies of natural hormones, or they may be slightly different.

HYBRID Usually refers to a natural or near-natural cross of two parent organisms that results in life forms with fairly predictable properties, as determined by the features of each parent.

HYDROPONICS The science of growing plants in a soil-free environment. Usually under glass with an *ad libitum* supply of water and minerals. The size of the resulting fruit and vegetables may be large, or even grotesque, but may be disappointing in flavour.

LISTERIA MONOCYTOGENES A bacterium commonly found in soil and the environment, but becoming dangerous after it has increased in numbers during refrigerated storage of food. Is also fairly resistant to heat. Can cause the illness listeriosis, resulting in miscarriage in pregnant women, or meningitis, particularly in elderly and ill people.

MICROORGANISM Any tiny organism, including viruses, bacteria, fungi, plankton.

MUTATION Damage or change in focal areas of DNA, with the production of different substances, usually to the detriment of the organism.

ORGANOPHOSPHOROUS COMPOUNDS 'Fashionable' insecticides that act on the insect's nervous system. Persist for less time than DDT

but can be dangerous to human handlers and can be environmentally
damaging, especially to marine life.

PHYTOPLANKTON Tiny marine organisms that use sunlight to synthesise the green pigment chlorophyll and some polyunsaturated fatty
acids. They provide the starting point for the natural food chain in
the sea.

RUMINANTS Mammals such as sheep, cattle, goats and deer that can
digest cellulose and related materials through bacterial activity in a
stomach chamber (the rumen), after which the food is regurgitated
and swallowed again.

SALMONELLA More than 2000 different species of bacteria found
naturally in mammals, birds and reptiles with secondary colonisation
of almost any food or organism. Named after an American, Dr Salmon, and capable of causing two types of disease in man. One is
enteric fever spread from person to person via food or water and is
a severe generalised infection of the body. The other is a type of food
poisoning from contaminated food, especially poultry meat and eggs.

SELECTION Artificial or natural processes that favour the survival and
then multiplication of certain organisms at the expense of others.

SPOILAGE The process whereby food goes bad, and often results from
the growth of bacteria such as *Pseudomonas*. Although such bacteria
may be harmless in themselves, they alert consumers to the possibility
of dangerous organisms.

SPORES Tiny, hard, inert, globular derivatives of bacteria formed
under inclement conditions to enable survival. Not dangerous in
themselves, but can return to growing bacteria that may be dangerous:
for example, *Clostridium botulinum* spores in canned food.

SOUS VIDE A risky procedure whereby meals are cooked, chilled and
stored in a vacuum for weeks, prior to reheating and eating.

ULTRAVIOLET LIGHT Shortwave solar irradiation liable to cause
mutations in cells, and even ultimately skin cancer.

VACCINE A man-made product that protects against a particular infectious disease.

VIRUS A tiny infectious agent containing DNA or RNA, which requires
a cell in which to multiply.

Further reading

Body, Sir Richard. *Our Food, Our Land: Why contemporary farming practices must change.* Rider, London, 1991.

Cannon, Geoffrey. *The Politics of Food.* Century, London, 1987.

Cannon, Geoffrey. *Food and Health: The experts agree.* Consumers Association, London (in press).

Consumers Association. *Understanding Additives.* Hodder and Stoughton, London, 1988.

Eastwood, Martin, Edwards, Christine and Parry, Doreen. *Human Nutrition: A continuing debate.* Chapman and Hall, London, 1992.

Erdmann, Robert and Jones, Meirion. *Fats, Nutrition and Health.* Thorsons, Wellingborough, UK, 1990.

Gaman, P.M. and Sherrington, K.B. *The Science of Food,* Third Edition. Pergamon, Oxford, UK, 1987.

Hughes, Christopher. *The Additives Guide.* John Wiley, Chichester, UK, 1987.

Lacey, Richard. *Unfit for Human Consumption.* Souvenir Press, London, 1990.

Lampkin, Nicholas. *Organic Farming.* Farming Press Books, Ipswich, UK, 1990.

Rifkin, Jeremy. *Beyond Beef: The rise and fall of the cattle culture.* Penguin, New York, 1992.

Walker, Caroline and Cannon, Geoffrey. *The Food Scandal.* Century, London, 1984.

Index